世界建筑 14
World Architecture 14

Art Museum
艺术博物馆 II

佳图文化 编

华南理工大学出版社
·广州·

图书在版编目（CIP）数据

世界建筑．14，艺术博物馆．2：英文/佳图文化编．—广州：华南理工大学出版社，2014.3
ISBN 978-7-5623-4080-5

Ⅰ．①世… Ⅱ．①佳… Ⅲ．①艺术—博物馆—建筑设计—作品集—世界—英文 Ⅳ．①TU206

中国版本图书馆CIP数据核字（2013）第243099号

世界建筑14：艺术博物馆Ⅱ
佳图文化 编

出 版 人：韩中伟
出版发行：华南理工大学出版社
（广州五山华南理工大学17号楼，邮编510640）
http://www.scutpress.com.cn　E-mail: scutc13@scut.edu.cn
营销部电话：020-87113487　87111048（传真）
策划编辑：赖淑华
责任编辑：骆　婷　黄丽谊
印 刷 者：广州市中天彩色印刷有限公司
开　　本：889mm×1194mm　1/16　印张：17
成品尺寸：245mm×325mm
版　　次：2014年3月第1版　2014年3月第1次印刷
定　　价：298.00元

版权所有　盗版必究　　印装差错　负责调换

Preface

Art centers, museums and other cultural buildings are important architectures for art appreciation or education purpose. In this new volume, we've selected some excellent projects to interpret the design skills of this kind of architecture. The book covers completed art buildings such as art galleries, museums, cultural centers, etc. around the world—we aim to include projects which are either of top quality or interesting, or ideally both. With professional design materials, high-resolution photographs and clear descriptions, it well introduces every project from all the aspects that will surely bring some inspirations to the readers.

Contents

Museum

National Museum of the Marine Corps — 002

Natural History Museum of Utah — 010

New Acropolis Museum — 018

Gran Museo del Mundo Maya de Mérida — 028

Renovation and Expansion of the Isabella Stewart Gardner Museum — 036

NASCAR Hall of Fame — 052

Exhibition Center

Cultural Center of EU Space Technologies — 060

Indiana Convention Center Expansion — 076

The Barnes Foundation — 082

Pavilion 4 — 094

Cultural Activity Building

Dalian International Conference Centre — 106

BRG Neusiedl am See — 118

Universidad Politecnica de Valencia Expansion —130

Comba Tentes Educational Center —142

Southland Christian Church —154

Thebarton Community Centre —160

Le Temps Machine —166

Auditorium del Parco —176

LA FABRIQUE, laboratoire(s) artistique(s) et centre culturel à Nantes —184

Birmingham Ormiston Academy —190

Art Building

Théâtre 95 —202

Theater De Nieuwe Kolk, Assen —216

New Theatre in Montalto di Castro, Italy —222

Opera House and Pop Music Stage Enschede —234

Pier K Theatre and Arts Centre —240

Hamer Hall —248

PALOMA Contemporary music venue at Nîmes Métropole —260

Museum

Education
Art Collection
Cultural Heritage

Keywords

Symbol
Angled
Flag-raising

National Museum of the Marine Corps

Location: Quantico, Virginia
Architect: Fentress Architect
Client: Marine Corps Heritage Foundation
Size: 11,150 m²

First prize winner in a national design competition, Curtis Fentress' design for the National Museum of the Marine Corps was honored with 20 awards in its first 20 months. Since President George W. Bush opened the museum in 2007, it has been visited by over one and a half million people, earning a reputation as a top cultural destination in the United States.

In seeking a symbol to illustrate the culture and history of the United States Marine Corps, Fentress Architects turned to an image embedded in the American soul since the final months of World War II: the flag-raising at Iwo Jima. This icon of determination, bravery and camaraderie inspired a winning competition design for a museum to serve as exhibition center, educational resource, and gathering place for a community much wider than Corps members alone. Through abstraction, the famous Joe Rosenthal photograph was translated into an angled, 210-foot stainless steel-clad mast that pierces a conical glass and steel skylight, giving the National Museum of the Marine Corps instant landmark status. Visible from Interstate 95, this bold form tops a circular building with clean lines expressed in simple, strong materials: cast-in-place concrete, steel, glass and stone. A broad plaza and entry create a natural link to the surrounding Semper Fidelis Memorial Park. The skylight also functions as a ceiling plane for the building's central gallery, where an efficient radial plan extends to house galleries, classrooms, visitor services, and restaurants. The museum's core is not just a connector. Eloquent quotations engraved in the inner walls speak of courage and commitment, while a terrazzo floor carries the pattern of a globe, representing

Master Plan

003

the land, sea and sky by which Marines have fought. Airplanes are suspended from the skylight, and other military transport vehicles are installed below, leading to immersion galleries. The circular form of the museum building also aids sustainable strategies. Part of the structure is embedded in earthen beams and a vegetated roof, which help regulate gallery temperature to conserve art and artifacts, provide acoustic buffering, and, in conjunction with the air flow, cool the building.

A nondenominational chapel sits atop a hill nearby, where visitors can reflect and worship. The wood, stone and glass-walled chapel is intended to host intimate events such as weddings, memorials and dedications.

"Military museums are not about war," says Curtis Fentress. "Instead, they celebrate the importance of maintaining peace and protecting citizens. At the National Museum of the Marine Corps, we mindfully searched for the ultimate icon to symbolize the end of war."

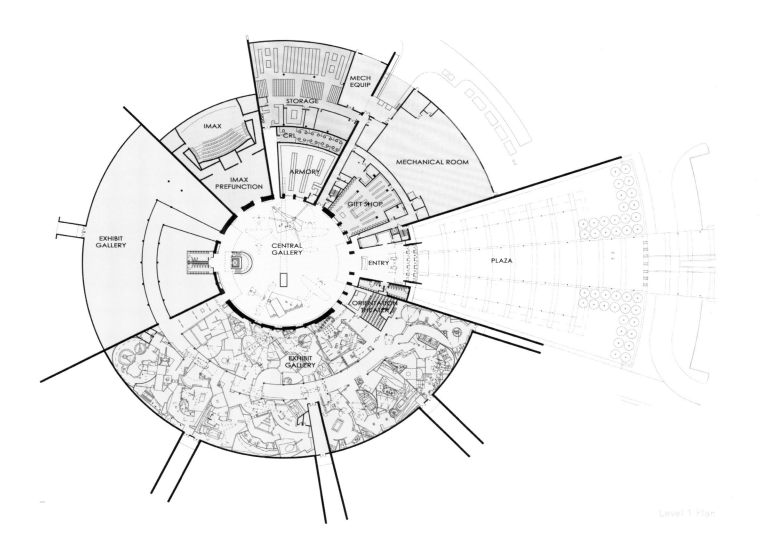

Level 1 Plan

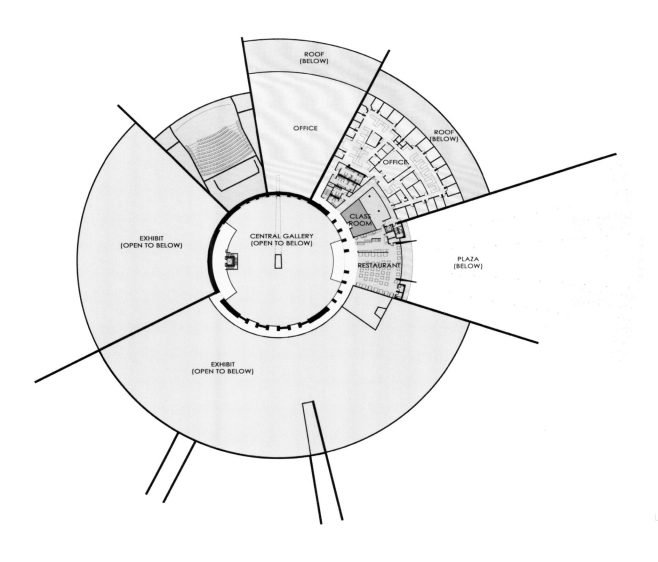

Level 2 Plan

005

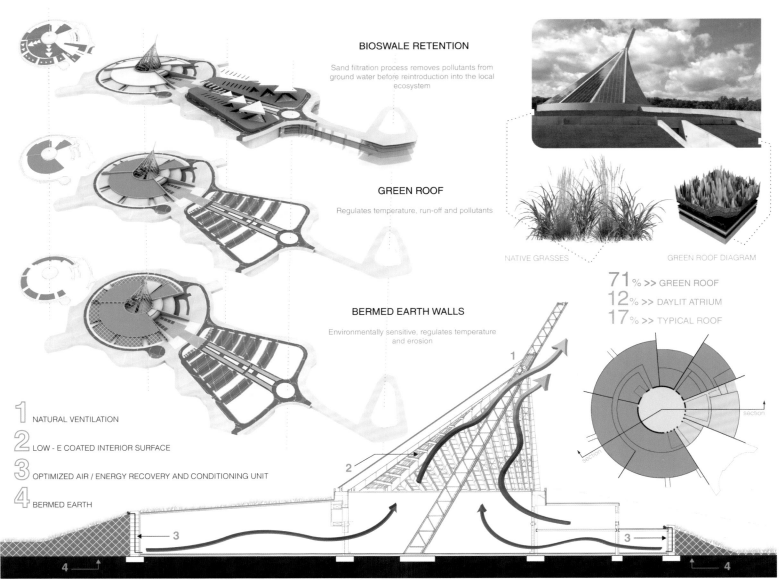

BIOSWALE RETENTION
Sand filtration process removes pollutants from ground water before reintroduction into the local ecosystem

GREEN ROOF
Regulates temperature, run-off and pollutants

BERMED EARTH WALLS
Environmentally sensitive, regulates temperature and erosion

NATIVE GRASSES   GREEN ROOF DIAGRAM

71% >> GREEN ROOF
12% >> DAYLIT ATRIUM
17% >> TYPICAL ROOF

1. NATURAL VENTILATION
2. LOW - E COATED INTERIOR SURFACE
3. OPTIMIZED AIR / ENERGY RECOVERY AND CONDITIONING UNIT
4. BERMED EARTH

Green Diagram

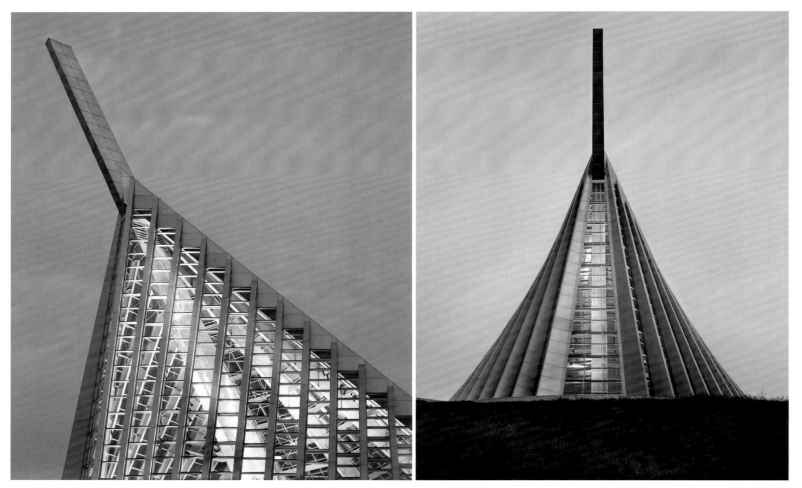

006 | Art Museum

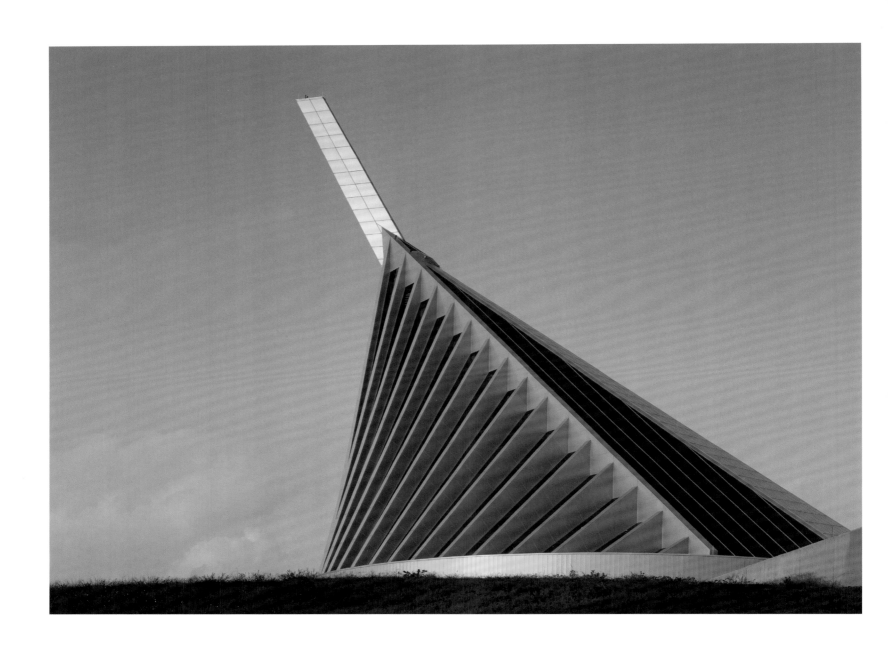

Keywords

Repository
Natural
Human

Natural History Museum of Utah

Location: Salt Lake City, Utah, United States
Architectural Design: ennead architects

The Natural History Museum of Utah (NHMU) is a museum located at the Rio Tinto Center on the campus of the University of Utah in Salt Lake City, Utah, United States. The museum shows exhibits of natural history subjects, specifically about Utah's natural history. The mission of the museum is to illuminate the natural world and the place of humans within it.

The museum's exhibit areas occupied almost 23,000 square feet (2,100 m^2) on the first and second floors of the George Thomas Building, located on the University of Utah campus. The exhibits targeted three broad areas of the natural sciences: geology/paleontology, anthropology, and biology.

The Museum is part of the academic life of the University of Utah. The collections offer research opportunities and provide a learning laboratory for students. Museum programs expose students to many aspects of museum studies: educational outreach, exhibit design and fabrication development, public relations, and curriculum development.

The Museum is a repository for collections that were accumulated by the University's departments of Anthropology, Biology, and Geology. The collections are held in trust for faculty, graduate students, and undergraduates who have access to the collections for research and teaching purposes.

In-service training is offered by the Utah Museum of Natural History Education Department; university credit can be earned with these courses, leading to salary lane changes for public school teachers. These courses are coordinated with the Academic Outreach and Continuing Education and the Department of Teaching and Learning. As the founder of the University's Genetic Science Learning Center, the Museum continues to partner in its teacher training program.

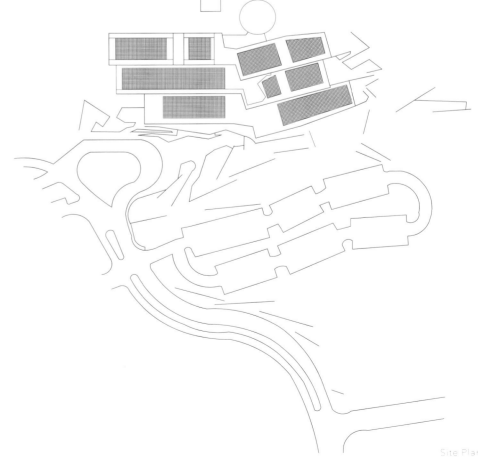

Site Plan

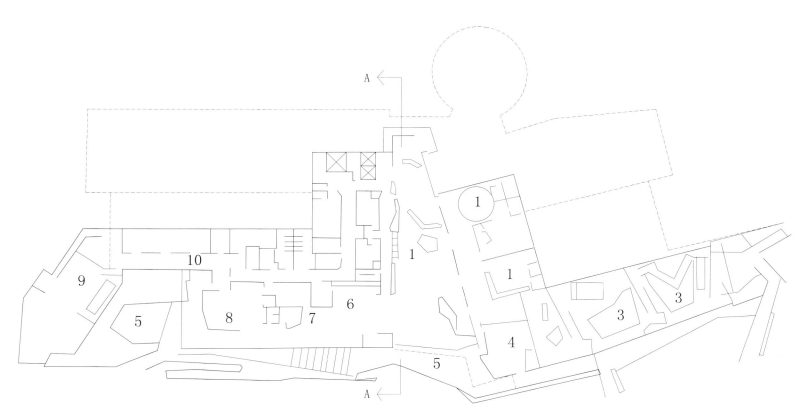

Second Level Floor Plan

1 RECEPTION
2 CANYON
3 PERMANENT EXHIBITS
4 PALEONTOLOGY PREPARATION LAB
5 TERRACE
6 MUSEUM STORE
7 CAFE
8 COMMUNITY MEETING ROOM
9 LOADING DOCK
10 SUPPORT SPACES

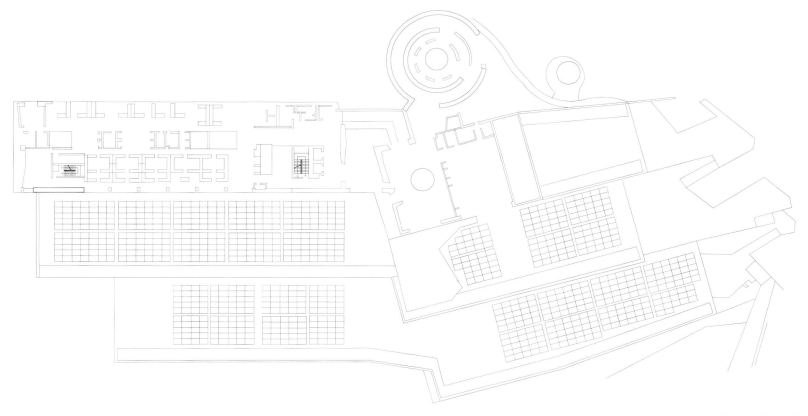

Fifth Level Floor Plan

017

Keywords

Story of Life
Utmost Simplicity
Circulation

New Acropolis Museum

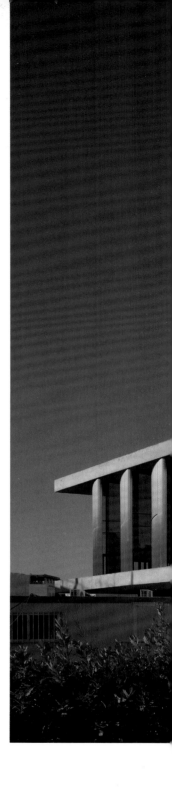

Location: Athens, Greece
Architectual Design: Bernard Tschumi Architects, New York and Paris
Associate Architect: Michael Photiadis ARSY Ltd., Athens
Structure: ADK and ARUP, New York
Mechanical and Electrical: MMB Study Group S.A. and ARUP, New York
Civil: Michanniki Geostatiki and ARUP, New York
Acoustics: Theodore Timagenis
Lighting: ARUP, London
General Contractor: Aktor
Size: 21,000 m² total; 8,000 m² of exhibition space
Completion: 2003-2009
Photography:Christian Richters and Peter Mauss/Esto

Program

With 8,000 m² of exhibition space and a full range of visitor amenities, the Acropolis Museum tells the story of life on the Athenian Acropolis and its surroundings by uniting collections formerly dispersed in multiple institutions, including the small Acropolis Museum built in the 19th century.

The rich collections provide visitors with a comprehensive picture of the human presence on the Acropolis, from pre-historic times through late antiquity. Integral to this program is the display of an archaeological excavation on the site: ruins from the 4th through 7th centuries A.D., left intact and protected beneath the building and made visible through the first floor. Other program facilities include a 200-seat auditorium.

Principal Design Features

Designed with spare horizontal lines and utmost simplicity, the museum is deliberately non-monumental, focusing the visitor's attention on extraordinary works of art. With the greatest possible clarity, the design translates programmatic requirements into architecture.

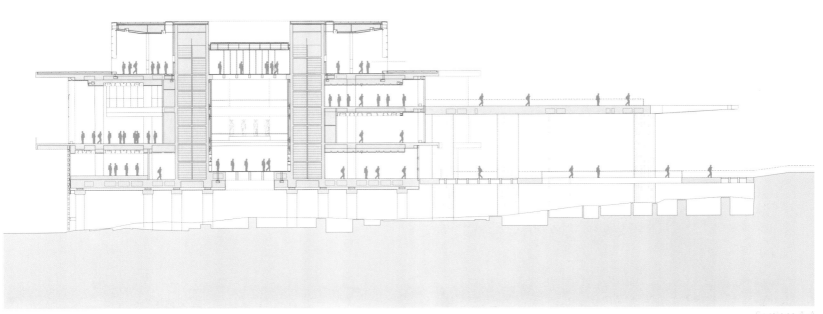

Sections A-A

Light

The collection consists primarily of works of sculpture, that many of their architectural pieces originally decorated the monuments making use of the Acropolis, so the building that exhibits them is a museum of ambient natural light. The use of various types of glass allows light to flood into the top-floor Parthenon Gallery, to filter through skylights into the archaic galleries, and to penetrate the core of the building, gently touching the archaeological excavation below the building.

Circulation

The collection is installed in chronological sequence, from pre-history through the late Roman period, but reaches its high point (literally and programmatically) with the Parthenon Frieze. The visitor's route is therefore a clear, three-dimensional loop. It goes up from the lobby via escalator to the double-height galleries for the Archaic period; upward again by escalator to the Parthenon Gallery; then back down to the Roman Empire galleries and out toward the Acropolis itself.

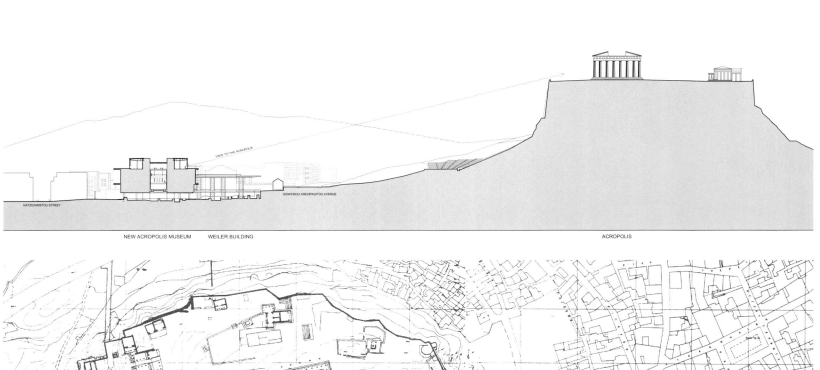

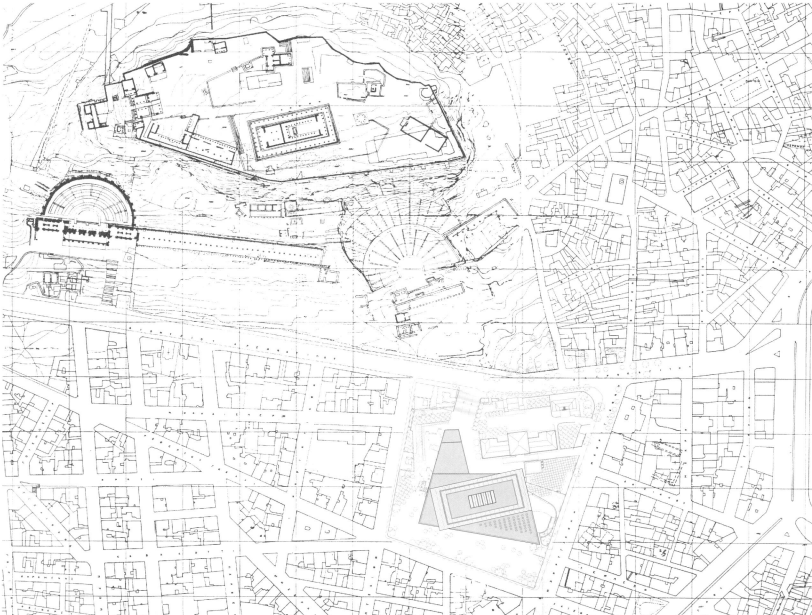

Master Plan

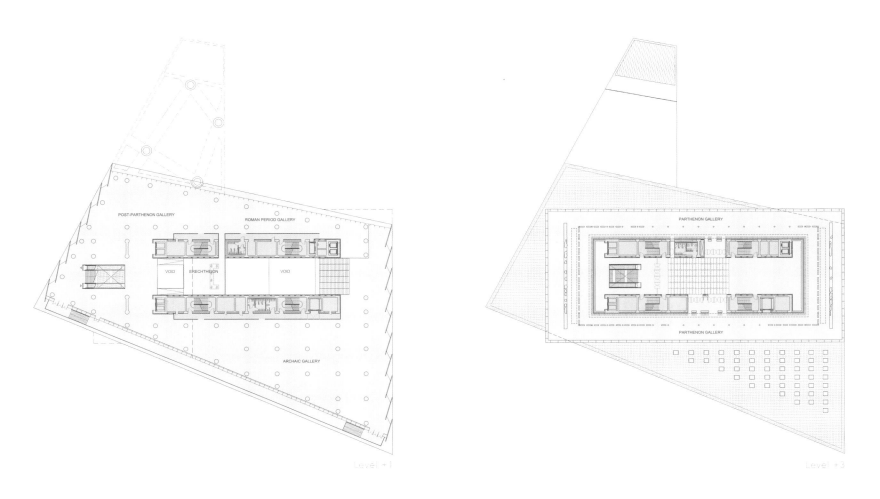

021

Organization

The Museum is conceived as a base, a middle zone and a top, taking its form from the archaeological excavation below and from the orientation of the top floor toward the Parthenon.

The base hovers over the excavation on more than 100 slender concrete pillars. This level contains the lobby, temporary exhibition spaces, museum store, and support facilities.

The middle (which is trapezoidal in plan) is a double-height space that soars to 10 m, accommodating the galleries from the Archaic to the late Roman period. A mezzanine features a bar and restaurant (with a public terrace looking out toward the Acropolis) and multimedia space.

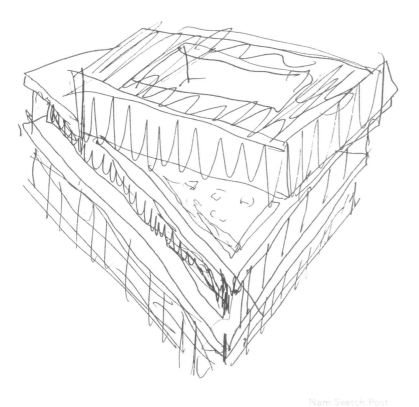

Nam Sketch Post

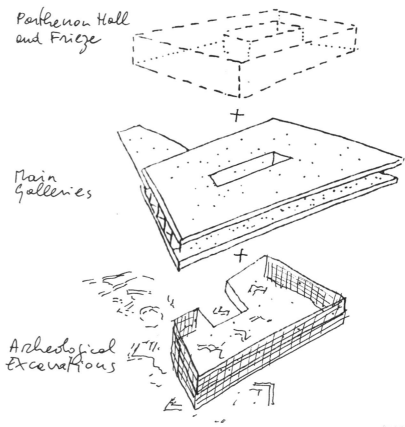

Parthenon Hall and Frieze

Main Galleries

Archeological Excavations

Axon

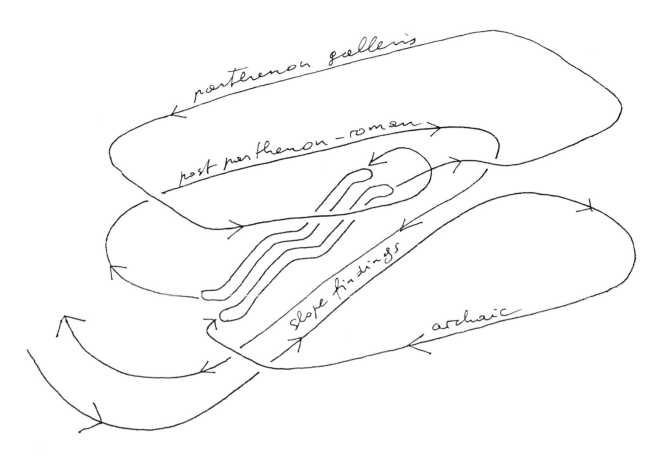

The top is the rectangular, glass-enclosed, skylit Parthenon Gallery, over 7 m high and with a floor space of over 2,050 m^2. It is shifted 23° from the rest of the building to orient that directly towards the Acropolis. Here the building's concrete core, which penetrates upward through all levels, becomes the surface on which the marble sculptures of the Parthenon Frieze are mounted. The core allows natural light to pass down to the Caryatids on the level below.

Keywords

Contemporary Expression
Sustainability
Multi Purpose

Gran Museo del Mundo Maya de Mérida

Location: Mérida, Yucatán, Mexico
Architectual Design: 4A Arquitectos
Design Team: Ricardo Combaluzier, Enrique Duarte, William Ramírez, Josefina Rivas
Collaborators: Luis De La Rosa, Alma Villicaña, Carlos Guardián, Mariana Farfán, Julio Rosas, Aída Ordóñez, Fabián Rosas, Ricardo Combaluzier
Area: 22,600 m²
Photography: David Cervera, Alessandra Ortíz, Hector Velasco, Tamara Uribe, Rocío Rojo

The Gran Museo del Mundo Maya is a building with a contemporary expression about what the Mayans worshiped rather than the Mayans built, in this search the architects found a recurrent symbol, a key element in the cosmic vision of Mayan Culture: Ceiba, the sacred tree, whose roots penetrate and conform the underworld, the trunk's level lays down where life and daily activities take place underneath the shade of its frond which spreads its branches up to the sky and human transcendence.

With this concept of the world's creation up from three stones and the Ceiba tree, the architects present the architectural design integrating the program needs and required spaces for the different functional activities, the structural design concept that gives physical bearing to the building and to the other infrastructure engineerings shared out for nurturing and supporting all the museum's areas. Museum collections, transit cellars, research and study areas of the great archaeological acquisition and a 260 parking spaces area is located at the "Ceiba's roots" level.

At the "Ceiba's trunk" level, up the perron, the main lobby, ticket offices, personal belongings kept area, 2,000 m² of permanent exhibition rooms and 600 m² for traveling exhibitions, public relations office, childcare center, restaurant with terrace, souvenir shop and a terrace bar. Executive and administrative offices are located inside the "Ceiba's frond", so are the high-definition large format cinema which includes performing arts facilities for various artistic and cultural activities as well as the multi-purpose hall. The architects understand sustainability as an integral part of any project and in this meaning the aspects considered are:

Photography: Tamara Uribe

Environment

Located at the heart of an important urban subcenter at north of Merida, the design raises seeking for natural air and light using passive systems to achieve energy benefits and environmental comfort. The main lobby covered and shaded by "La Ceiba" which holds a hollow core, joints the different floors of the building.

Society

The architects designed a comprehensive museum, pretending to make suitable space to every user: a twined ramp at the perron, a sidewalk-level elevator and another one in the parking lot, to yield universal access with equal dignity, Braille signaling, all facilities in corridors and restrooms for the elderly or handicapped people, rest areas while taking the tour, spaces for workers to improve their life quality, machinery and equipment rooms designed for both, machines and people who operate them, bestowing to workers the same importance as visitors.

Economy

Significant economic rationality guidelines are followed, displayed in the design of functional spaces with direct use of passive systems and the selection of materials and building systems that privileged participation of local and regional companies and their employees; and the optimization in infrastructure engineerings projects to achieve the best use of resources for the operation and maintenance.

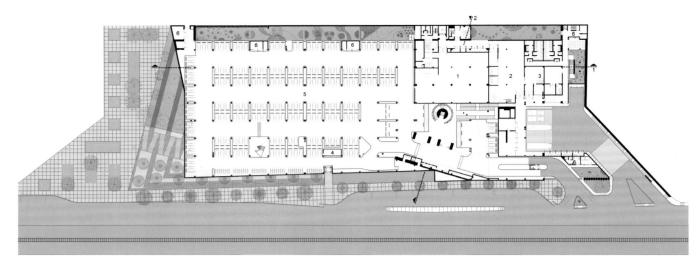

1. MUSEUM COLLECTIONS
2. TRANSIT COLLECTION AREA
3. STAFF AREA
4. MULTIMEDIA ROOM
5. PARKING
6. MECHANICAL EQUIPMENT ROOM
7. PERMANENT EXHIBITION ROOMS
8. TRAVELLING EXHIBITION ROOM
9. MAIN LOBBY
10. TICKET OFFICES
11. SOUVENIR SHOP
12. RESTAURANT
13. PUBLIC RELATIONS OFFICE
14. CHILDCARE CENTER
15. RESTAURANT TERRACE
16. FAMILY RESTROOMS
17. BATHROOMS FACILITIES
18. TERRACE BAR
19. HIGH-DEFINITION LARGE FORMAT CINEMA
20. CAFE
21. MULTI-PURPOSE HALL
22. ADMINISTRATIVE OFFICES

GROUND FLOOR

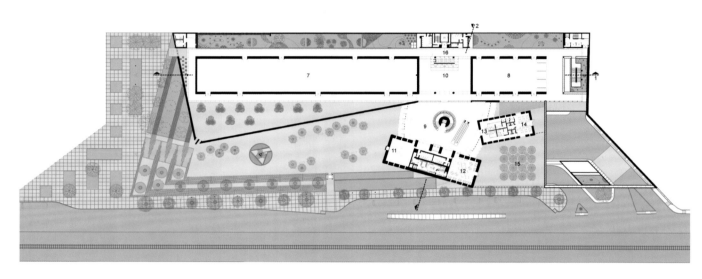

1. MUSEUM COLLECTIONS
2. TRANSIT COLLECTION AREA
3. STAFF AREA
4. MULTIMEDIA ROOM
5. PARKING
6. MECHANICAL EQUIPMENT ROOM
7. PERMANENT EXHIBITION ROOMS
8. TRAVELLING EXHIBITION ROOM
9. MAIN LOBBY
10. TICKET OFFICES
11. SOUVENIR SHOP
12. RESTAURANT
13. PUBLIC RELATIONS OFFICE
14. CHILDCARE CENTER
15. RESTAURANT TERRACE
16. FAMILY RESTROOMS
17. BATHROOMS FACILITIES
18. TERRACE BAR
19. HIGH-DEFINITION LARGE FORMAT CINEMA
20. CAFE
21. MULTI-PURPOSE HALL
22. ADMINISTRATIVE OFFICES

FIRST LEVEL

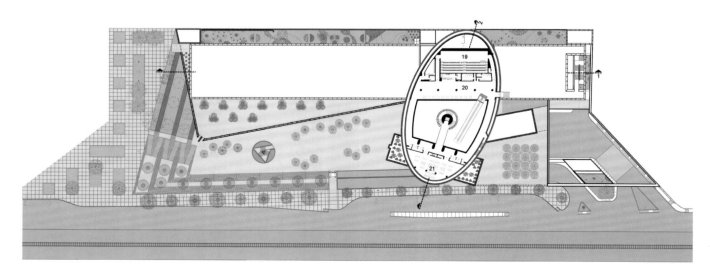

1. MUSEUM COLLECTIONS
2. TRANSIT COLLECTION AREA
3. STAFF AREA
4. MULTIMEDIA ROOM
5. PARKING
6. MECHANICAL EQUIPMENT ROOM
7. PERMANENT EXHIBITION ROOMS
8. TRAVELLING EXHIBITION ROOM
9. MAIN LOBBY
10. TICKET OFFICES
11. SOUVENIR SHOP
12. RESTAURANT
13. PUBLIC RELATIONS OFFICE
14. CHILDCARE CENTER
15. RESTAURANT TERRACE
16. FAMILY RESTROOMS
17. BATHROOMS FACILITIES
18. TERRACE BAR
19. HIGH-DEFINITION LARGE FORMAT CINEMA
20. CAFE
21. MULTI-PURPOSE HALL
22. ADMINISTRATIVE OFFICES

SECOND LEVEL

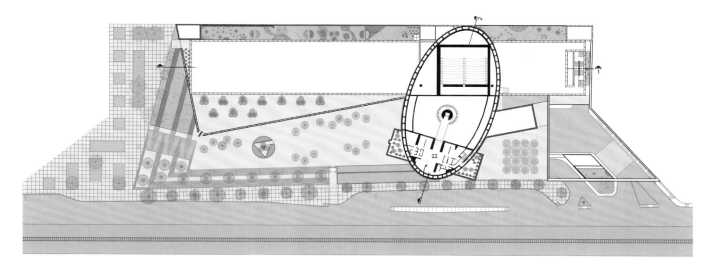

1. MUSEUM COLLECTIONS	7. PERMANENT EXHIBITION ROOMS	13. PUBLIC RELATIONS OFFICE	19. HIGH-DEFINITION LARGE FORMAT CINEMA
2. TRANSIT COLLECTION AREA	8. TRAVELLING EXHIBITION ROOM	14. CHILDCARE CENTER	20. CAFE
3. STAFF AREA	9. MAIN LOBBY	15. RESTAURANT TERRACE	21. MULTI-PURPOSE HALL
4. MULTIMEDIA ROOM	10. TICKET OFFICES	16. FAMILY RESTROOMS	22. ADMINISTRATIVE OFFICES
5. PARKING	11. SOUVENIR SHOP	17. BATHROOMS FACILITIES	
6. MECHANICAL EQUIPMENT ROOM	12. RESTAURANT	18. TERRACE BAR	

THIRD LEVEL

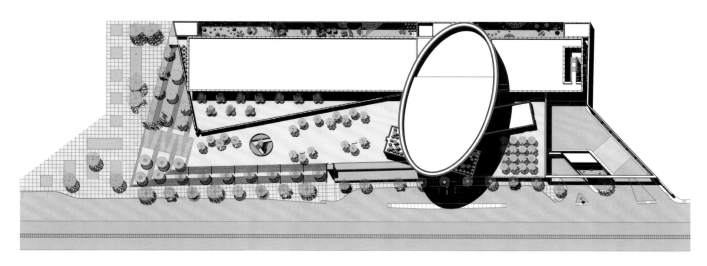

SITE PLAN

Photography: David Cervera

Photography: David Cervera

Photography: David Cervera

Photography: David Cervera Photography: Hector Velasco

Photography: Hector Velasco

Photography: Tamara Uribe

Photography: David Cervera

Photography: David Cervera

Photography: Hector Velasco

Photography: Rocío Rojo

Photography: Hector Velasco

Keywords

Uninterrupted Views
Historic Purpose
Sustainable Design

Renovation and Expansion of the Isabella Stewart Gardner Museum

Location: Boston, USA
Client: Isabella Stewart Gardner Museum
Architectual Design: Renzo Piano Building Workshop
In Collaboration With: Stantec - Burt Hill (Boston)
Design Team: E.Baglietto (partner in charge) with M.Aloisini, I.Ceccherini V.Grassi, S.Ishida (partner), Y.Kim, M.Liepmann, M.Neri, K. Schorn, T.Stewart, O.Teke and E.Moore; G.Langasco (CAD Operator); F.Cappellini, A.Marazzi, F.Terranova (models)
Consultants: Buro Happold (structure and services); Front (façade consultant); Arup (lighting); Nagata Acoustics (acoustics); Stuart-Lynn Company (cost consultant); Paratus Group (project manager) CBT/Childs Bertman Tseckares (consulting architect for Schematic & Design Development)
Total Floor Area of the New Wing: 6500 m²
Total Area of the Historic Museum Building: 5300 m²

The design of the Museum's new wing incorporates glass and natural light to create an open and welcoming entrance, as well as to provide uninterrupted views of the historic building and gardens. The building features four volumes clad in green pre-patinated copper and red brick that appear to "float" above the transparent first floor. Key features of the new wing are a cube-shaped performance hall and an adjustable height special exhibition gallery, which are the Museum's first purpose-built spaces to accommodate such functions.

Visitors enter the Museum through a new entrance facing Evans Way Park into the glass-enclosed Bekenstein Family Lobby. A new space, named the Richard E. Floor Living Room, welcomes the visitor in an intimate domestic-like setting where hosts, books, and touch screen monitors on easels offer information about Isabella Stewart Gardner, the collection and its unique installation, and the Museum's Artist-in-Residence Program.

Calderwood Hall, the Museum's new performance hall, is the largest space in the new wing at 557 m², and is designed in collaboration with acoustician Yasuhisa Toyota of Nagata Acoustics. With 300 seats configured in three balcony levels surrounding the central performing area on all four sides, the hall preserves the intimate experience that has long characterized the Gardner Museum's music program.

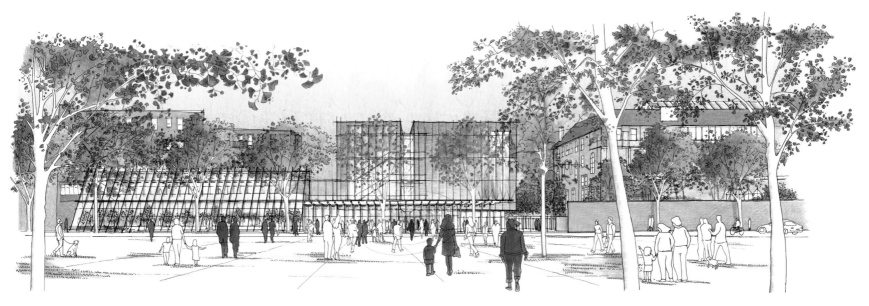

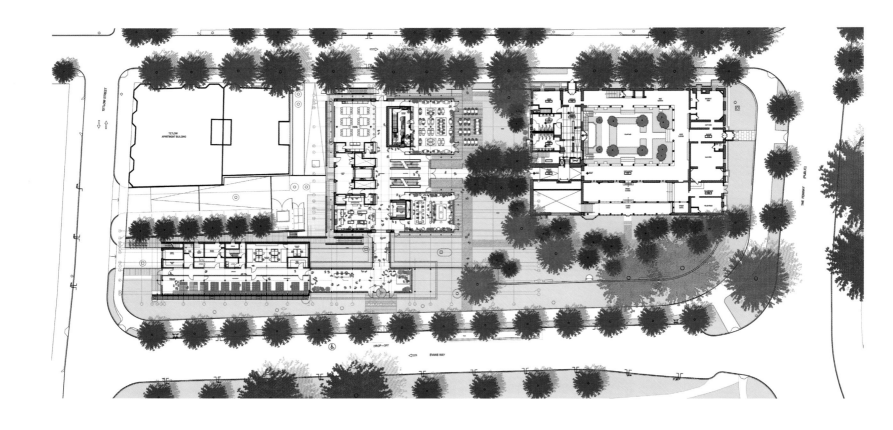

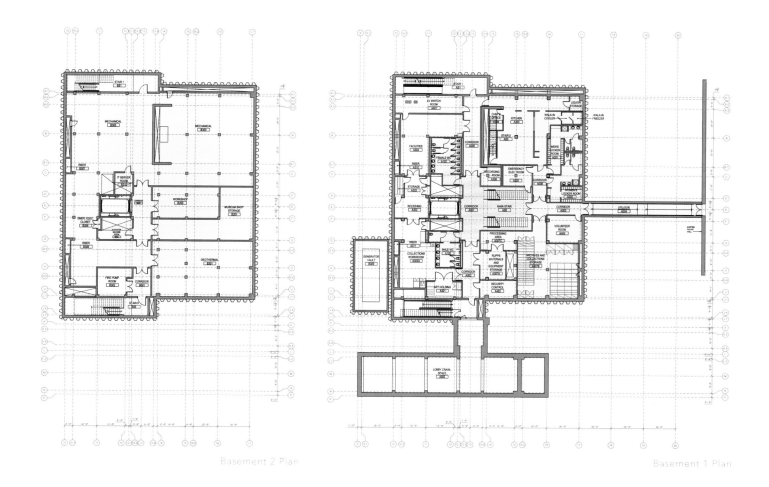

Basement 2 Plan

Basement 1 Plan

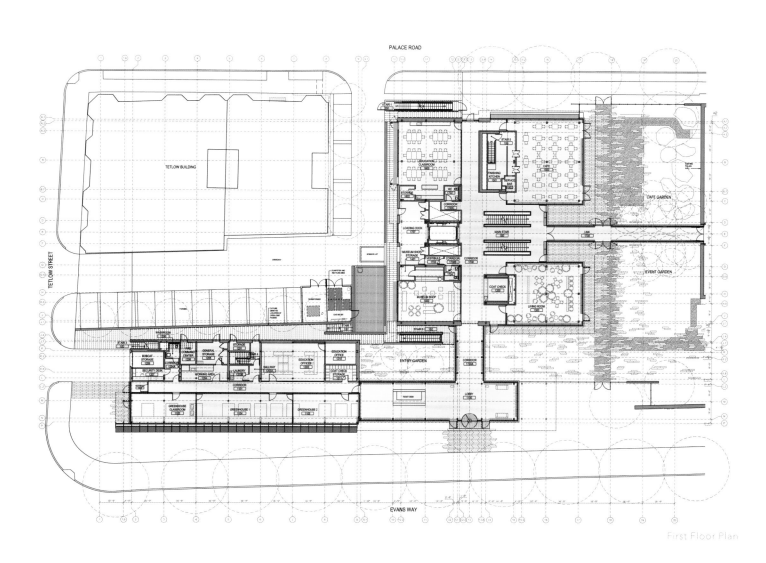

First Floor Plan

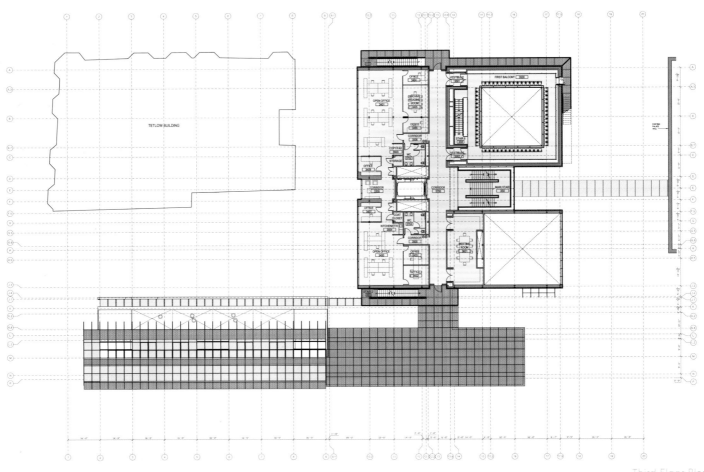

Third Floor Plan

040 | Art Museum

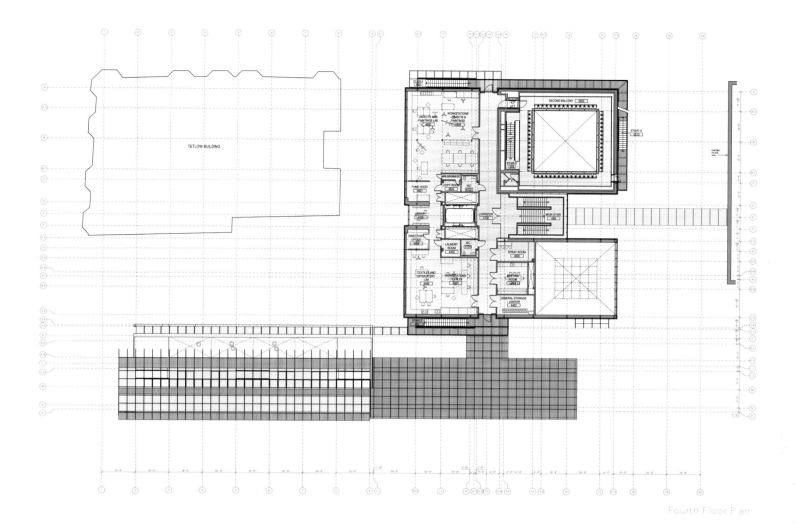

Fourth Floor Plan

The Special Exhibition Gallery, which will present three major exhibitions each year, is a flexible space featuring a retractable ceiling and a full wall of windows overlooking the historic Museum and the Monks Garden. The addition also houses working greenhouses, a landscape classroom and expanded outdoor garden spaces, two artist apartments, conservation labs, the Claire and John Bertucci Education Studio which will offer hands-on art workshops for students and families, a new store called Gift at the Gardner, and a new restaurant, Café G, with indoor and seasonal outdoor seats.

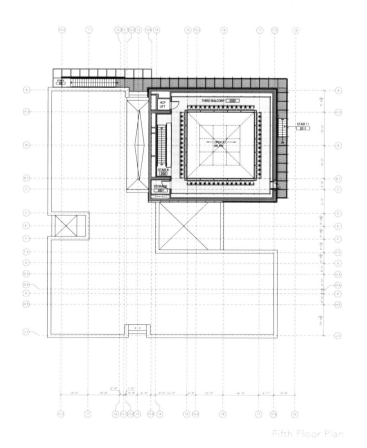

Fifth Floor Plan

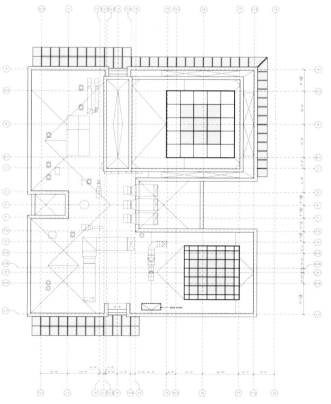

Roof Plan

"Isabella Gardner's Palace, with its treasured collection and inimitable installations, its verdant courtyard and mesmerizing corridors, will always be the focus of the Museum, but it could only remain so with the construction of a companion building. With housing for resident scholars and artists, labs for the conservation of the collection, and room for public assembly and school partners, the new wing frees up the historic building to fulfill its historic purpose," added Hawley. The Isabella Stewart Gardner Museum is seeking LEED gold certification by the United States Green Building Council. Primary components of the sustainable design are a geothermal well system, daylight harvesting, water-efficient landscaping techniques, and the use of local and regional materials, which reduces the environmental impact associated with transport.

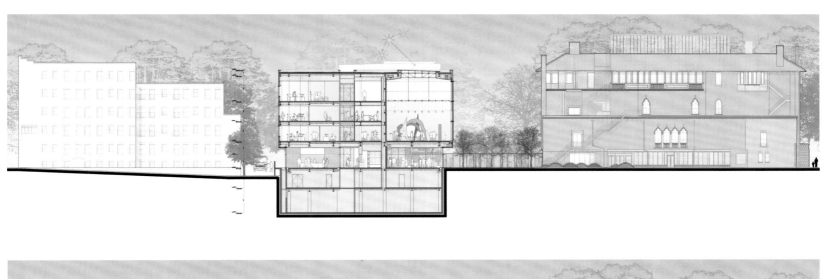
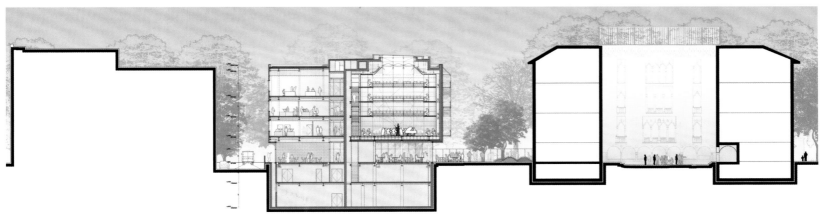

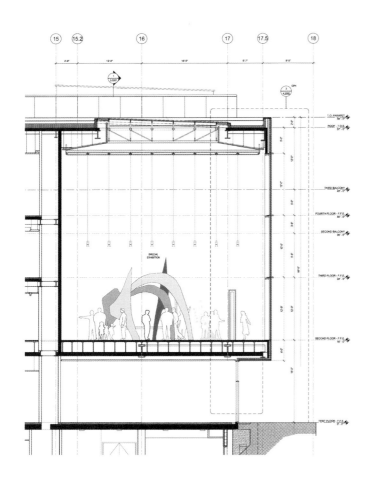
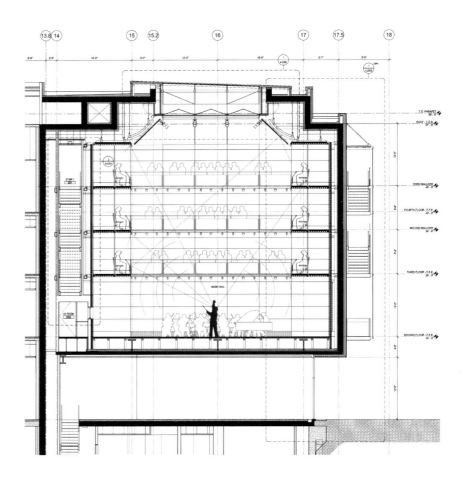

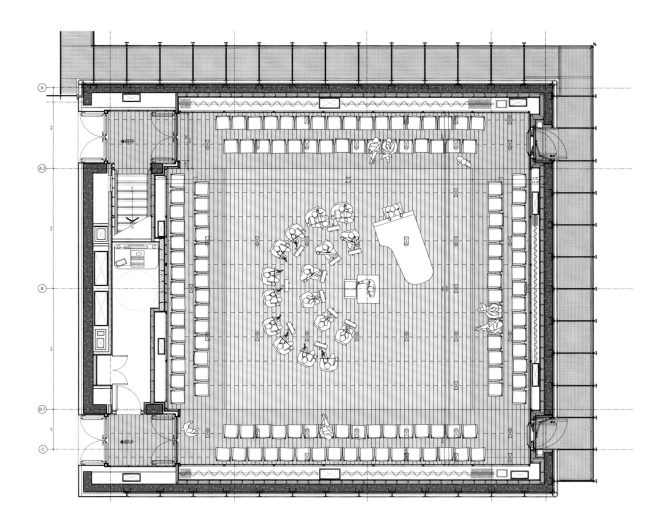
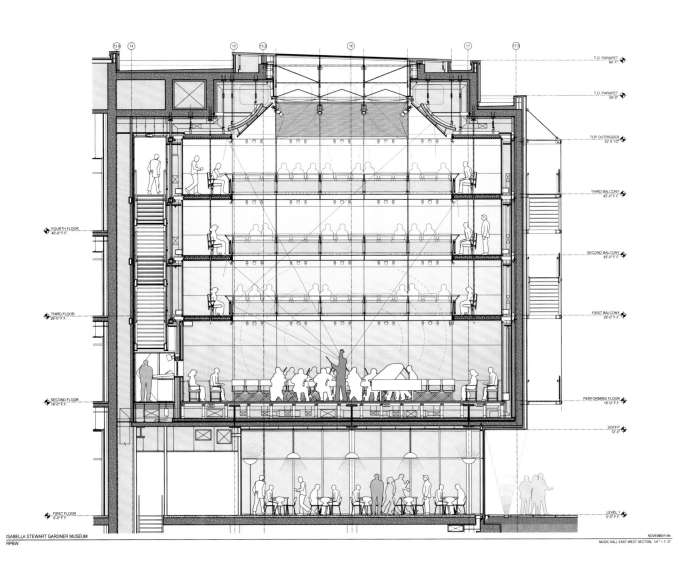

047

Keywords

Architectural Form
Speed and Spectacle
Gateway

NASCAR Hall of Fame

Location: Charlotte, North Carolina, USA
Architectual Design: Pei Cobb Freed & Partners
Project Area: 65,032 m^2
Photography: Peter Brentlinger, Paul Warchol

In approaching the challenge of designing a Hall of Fame for NASCAR, Pei Cobb Freed & Partners sought to capture the essential spirit of NASCAR and its sport in architectural form. In exploring the possibilities for expressing speed and spectacle, they were drawn to the arena of action, the racecourse, where fans and race teams come together each race week for the spectacle of race day. Curving and sloped forms are evocative not only in the dynamic and changing sinuous shape of the racetrack but also in the perception of speed, which is at the heart of the NASCAR spectacle.

An important part of Pei Cobb Freed & Partners's design strategy is to locate NASCAR Hall of Fame Plaza, a sweeping forecourt that welcomes visitors to the Hall of Fame, on the northern edge of the site, poised toward the pedestrian traffic and energy of uptown Charlotte.

The Hall of Fame consists of four basic elements. First, a large glazed oval shape forming a Great Hall serves as the symbolic core of the Hall of Fame and a primary orientation point for visitors. Second, a rectangular volume houses visitor services, including entry and exhibit space on the upper floors. Third, an expressed Hall of Honor is situated as an iconic element within the Great Hall. Finally, a broadcast studio, serving as an origination point for a variety of radio and television broadcasts by NASCAR's media partners, enlivens the Plaza.

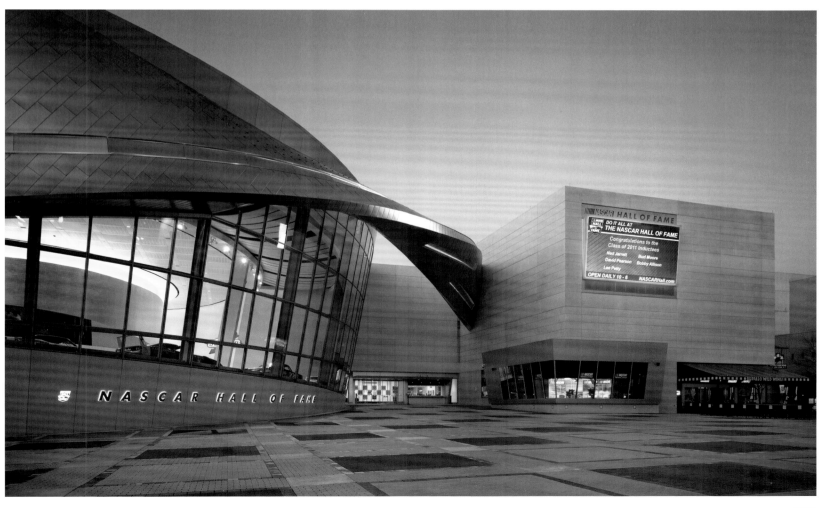

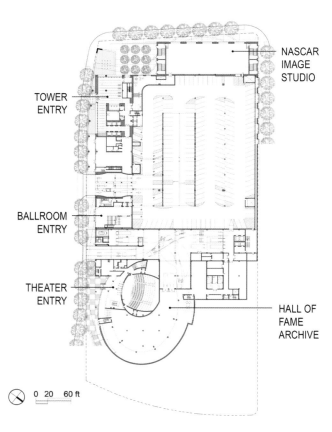

Tower Entry Level

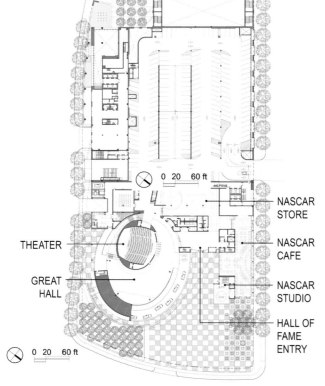

Plaza Level

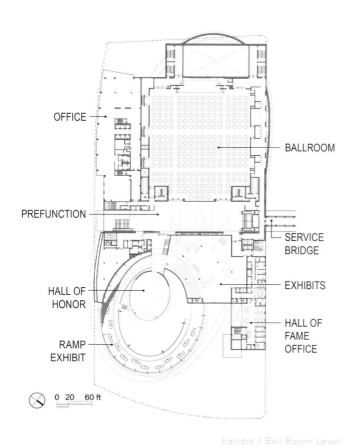

Exhibit / Ball Room Level

- OFFICE
- BALLROOM
- PREFUNCTION
- SERVICE BRIDGE
- EXHIBITS
- HALL OF HONOR
- HALL OF FAME OFFICE
- RAMP EXHIBIT

0 20 60 ft

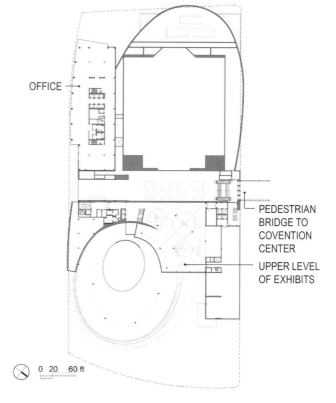

Upper Exhibit Level

- OFFICE
- PEDESTRIAN BRIDGE TO COVENTION CENTER
- UPPER LEVEL OF EXHIBITS

0 20 60 ft

The results of Pei Cobb Freed & Partners's explorations of speed and spectacle evolves into an architectural element they call the Ribbon, which envelops the varied program elements in a form that speaks to the imagery and spirit of NASCAR. Beginning as a curved, sloping exterior wall enclosing the building, the Ribbon twists in a free span over the main entry to form a welcoming canopy. Inspired by the dynamic quality of speed, captured in a second as a blur on film, the long, thin incisions in this metal skin are analogous to the blur of a car racing past the spectator at tremendous speed.

On the interior within the Great Hall, a signature element of a curved banked ramp leads visitors from the main floor to the exhibit levels above. The ramp contains a display of race cars frozen in a moment from a race, capturing in another way the speed and spectacle that are the essences of the sport.

NASCAR Tower is a twenty-storey office tower anchoring the southeast corner of the full city block development that is the Hall of Fame Complex. Located at the intersection of the Caldwell Street freeway interchange, the tower is designed as a gateway to the city and to blend with the iconic design of the Hall of Fame. The tower's form follows the lead of the Hall of Fame, consisting of a curvilinear metal and glass curtain wall contrasted with a rectangular precast concrete armature.

Exhibition Center

Modernity
Public Building
Exhibition Use

Keywords

Interlocking Rings
Monolithic Concrete Structure
Culturalisation of Space

Cultural Center of EU Space Technologies

Location: Vitanje, Slovenia
Client: KSEVT, Vitanje Community and Ministery of Culture
Architectual Design: Dekleva Gregoric Arhitekti + SADAR + VUGA + OFIS architects + Bevk Perovic Arhitekti
Site Area: 5,175 m^2
Area: 2,505 m^2
Photography: Tomaz Gregoric, Miran Kambic

The Cultural Center of European Space Technologies (KSEVT) will substantially supplement and emphasize the local cultural and social activities of the former Community Center in Vitanje, the town in Slovenia that was family home to Herman Potocnik Noordung, the first theoretician of space. The program includes additional cultural (exhibitions, events) and scientific activities (research, conferences) strongly connected to the phenomena of "culturalization of the space".

The building features a series of interlocking rings that lie on top of each other to create a continuous ramped structure. The design integrates two buildings in one: a local community centre with a circular multi-purpose hall and local library and the museum of Space Technologies with its exhibition and research areas. KSEVT will have a public significance and generate social, cultural, and scientific activities, with fixed and temporary exhibitions, conferences and club/study activities.

The concept design for the building of the KSEVT derives from the habitation wheel of the first geostationary space station described in Noordung's 1929 book titled "The Problem of Space Travel – The Rocket Motor". The main exhibition space circularly wraps the main round hall, which connects with the research spaces above through the round opening between the two spaces. It creates interaction between the program of local community and the scientific program of KSEVT.

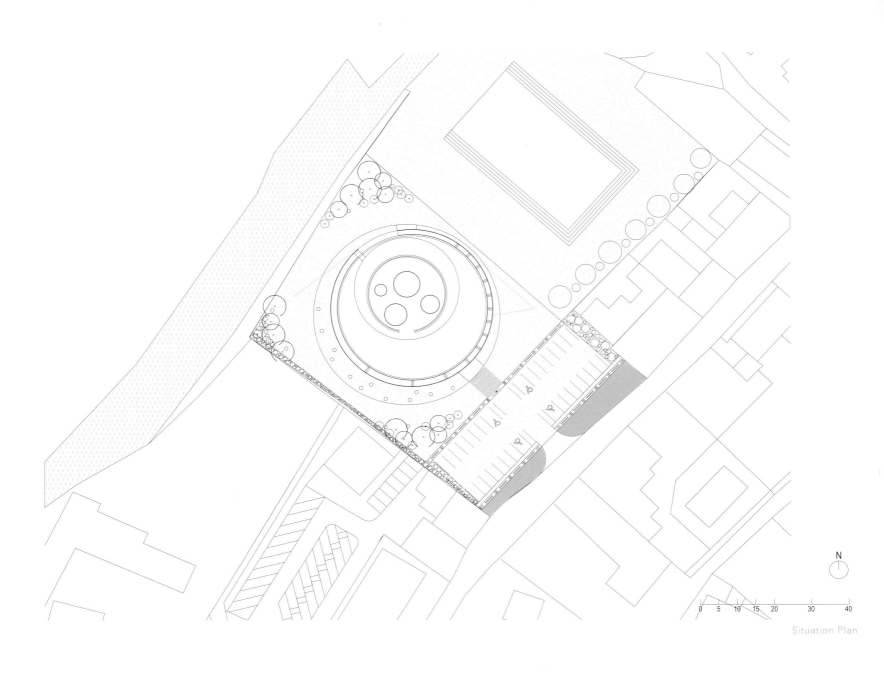

Situation Plan

062 | Art Museum

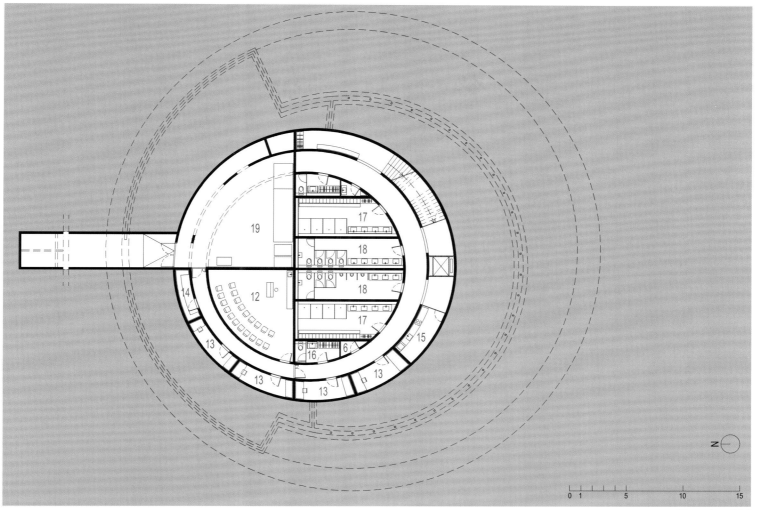

Floorplan -3.55

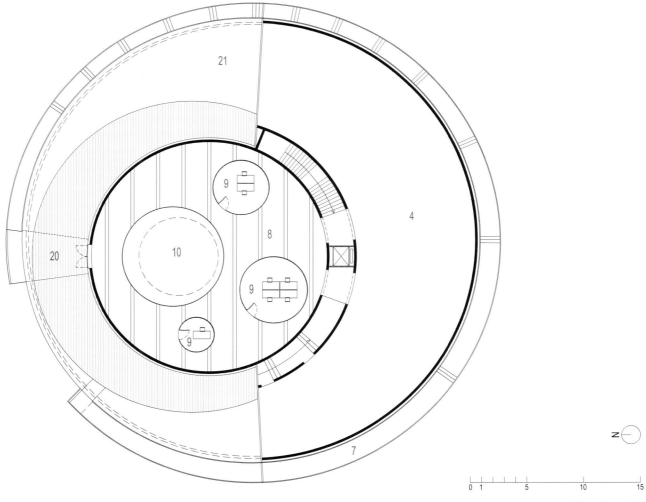

Floorplan +6.90

063

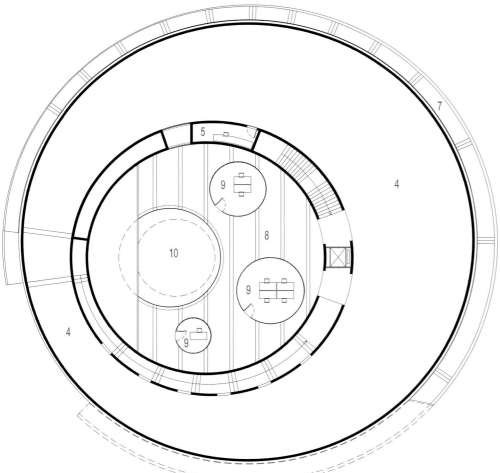

The building exceeds the size of the generic houses of Vitanje. With its volumetric and dynamic presence builds the relation with the main church that sits on a little hill in the middle of the town. In this way symbolically represents the Science & Culture counterpart to the Religion & Church.

Noordnung's space station was designed as a geostationary satellite out of three parts: a solar power station, an observatory and a habitable wheel. After several decades of pondering on the habitation of space, this idea remains to be the most revolutionary, yet not realized. The rotating habitable wheel, a circular construction setting up artificial gravity with the centrifugal force, is the best and at the same time a simple solution for long-term human habitation of weightlessness. Since we are not accustomed to that kind of condition, it exerts negative influence upon our body in the long run. A station in this orbit could also represent a perfect point of departure for longer spaceflights, considering that the Earth's force of attraction is still the greatest obstacle for that.

The building is a monolithic concrete structure, positioned freely between a main road on one side and a stream with a green hinterland on the other. The exterior and interior of the building are made of two low cylinders. The bottom one is larger and rises from the North to the South, while the upper cylinder is smaller and joins the larger one on the south while rising to the North. The bottom cylinder is supported by the transparent surface of the entrance glazing.

From the exterior, there is a dynamic effect between the cylinders, accentuated by the full glass rings around the building. The building appears to float and rotate on its southern and western sides towards the road. The entrenchment of the building into the surface on the other side gives a connection to its immediate surroundings. The spatial effects give the building the effect of artificial gravity from floatation and rotation. The building has two entrances—a main one to the central space from the square in front of the building on the south-eastern side and the northern entrance from the gravel surface above the stream.

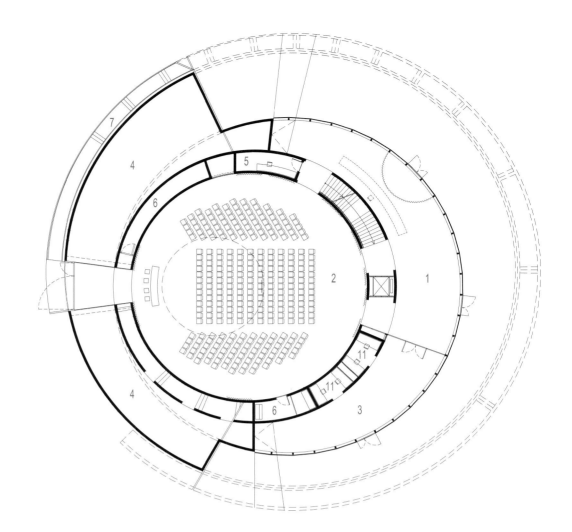

The main entrance covers the overhanging part of the bottom cylinder: one passes through a tight space past a circular vestibule and into the interior of the hall. The vestibule can be separated from the activities in the hall by a curtain. The entrance glazing can be completely opened and can connect the activities in the hall with the square. The circular hall for 300 people is surrounded on both sides by a semicircular ramp. This denotes the beginning of the exhibition area, continuing from here to the overhanging part of the larger cylinder.

On the west, there are smaller office areas along the ramp. Ascending this ramp also represents a transition from the bright space of the hall to the dark exhibition area. The vertical connection with a staircase and large elevator connects the exhibition area directly to the vestibule of the hall. The exhibition space continues through the landing between the elevator and the staircase to the smaller cylinder, the multi-purpose hall, and a raised auditorium above the hall. From here, one can observe the activity below. The smaller cylinder is concluded at the highest, northernmost portion with a club area devoted to researchers of the history of space technology, where they can focus on their work aside from the activities below.

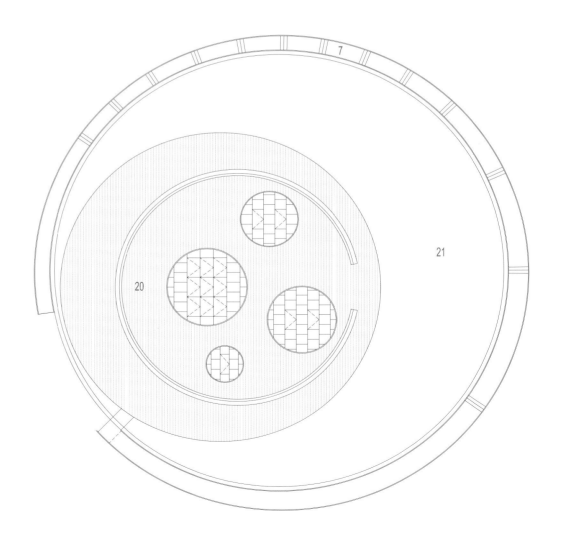

065

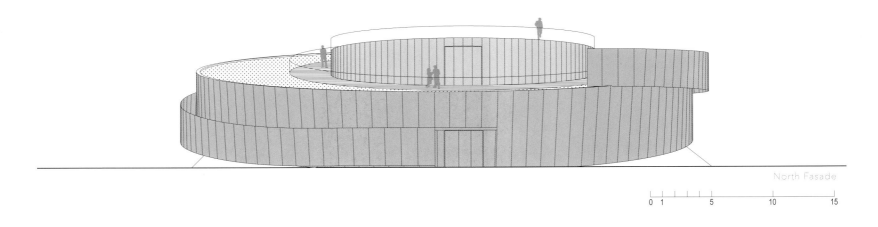

North Fasade

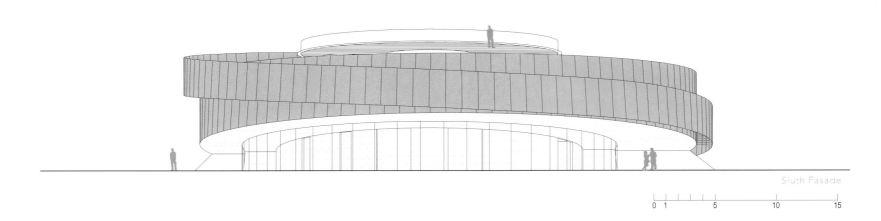

South Fasade

Besides special programme and location, the collaboration of four architectural offices in developing the project is also unique. The idea of collaboration was raised on the first meeting where investor invited the four offices to collaborate on internal competition – and office principals decided to actually do the project together. The idea concepts came out on serious of workshops, later project was shared in different stages of development between all offices.

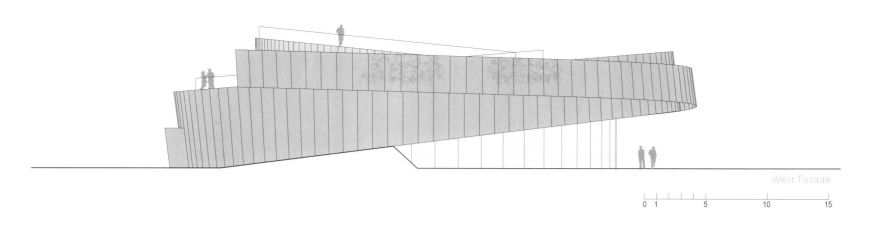

West Fasade
0 1 5 10 15

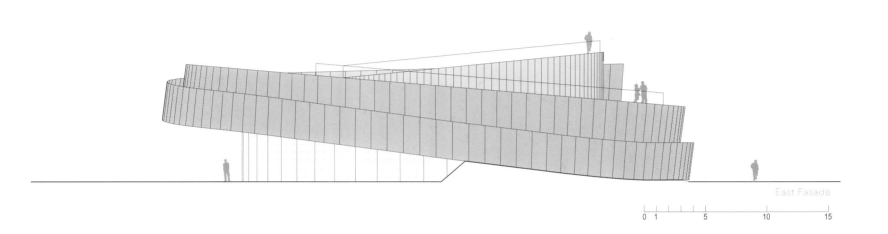

East Fasade
0 1 5 10 15

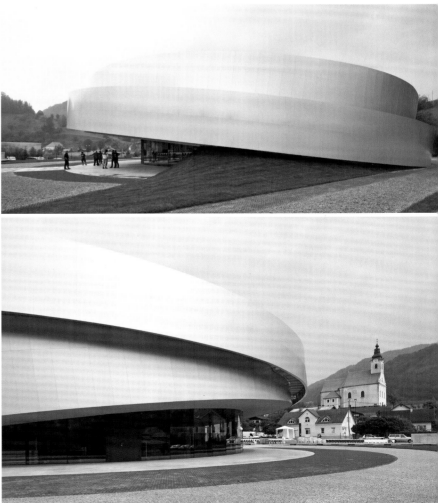

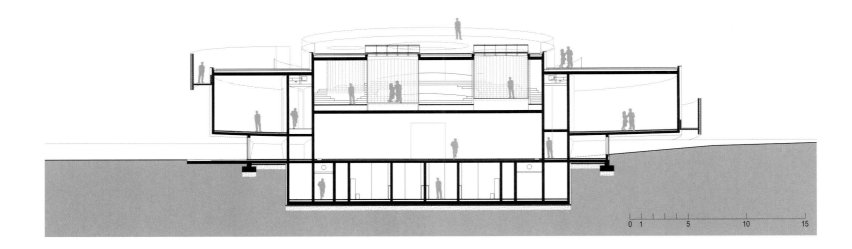

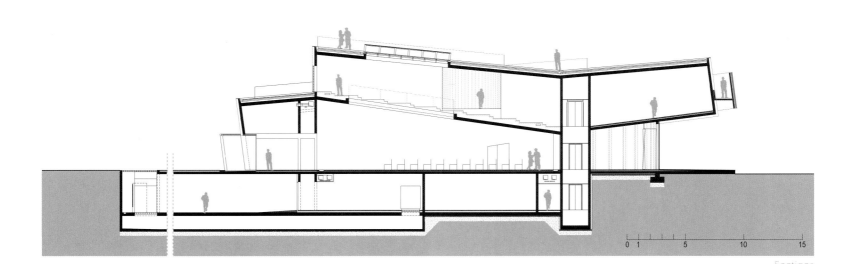

Sections

068 | Art Museum

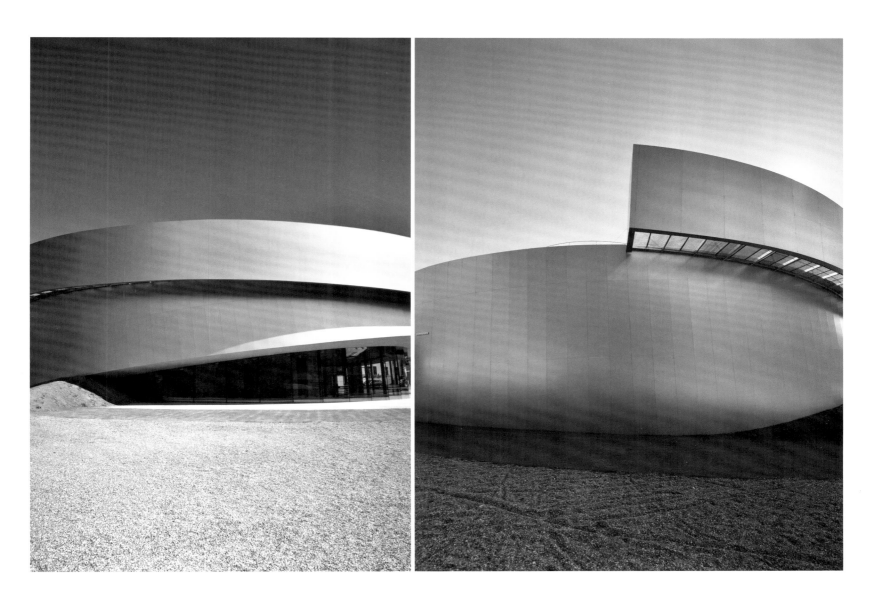

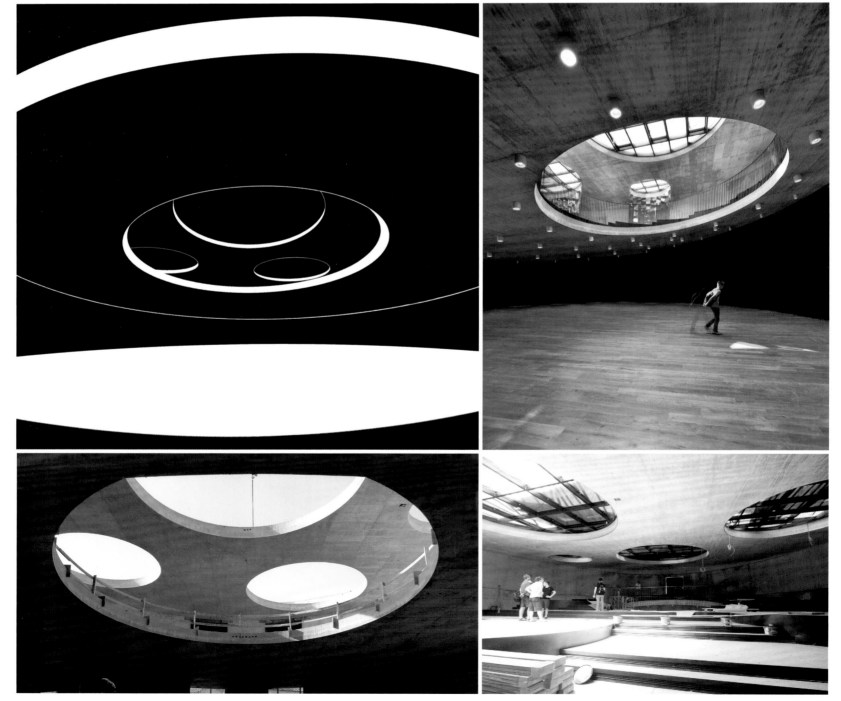

Keywords

Meeting Place
Climate-controlled Passageway
Clear Glass Cube

Indiana
Convention
Center Expansion

Location: Indianapolis, Indiana, United States
Architectual Design: RATIO Architects
Area: 65,000 m²
Completion: 2011
Photography: Bill Zbaren Photography

In 2006, the Indiana Stadium and Convention Building Authority selected RATIO Architects to create a major expansion to the Indiana Convention Center as a response to ongoing increased convention and trade show demand. The more than 65,032 m² expansion includes exhibition space, meeting rooms, and pre-function and support space – all within a tight urban site in the heart of Indianapolis, directly connected to 4,700 hotel rooms and within easy walking distance of restaurants, retail shops and at the head of a new pedestrian/event street.

The project also features a climate-controlled pedestrian passageway between the Indiana Convention Center and the city's new Lucas Oil Stadium. The 65 m pedestrian corridor includes a tunnel below highly-valued and heavily-used CSX and Amtrak rail lines that remain active throughout construction. The $275-million expansion of the Indiana Convention Center, together with facilities at the connected Lucas Oil Stadium, creates the 16th largest convention complex in the U.S., elevating Indianapolis to the top tier of mid-level convention cities. To mark its place as an Indiana facility, the Indiana Convention Center expansion is designed to recall the narrative story of the glacially-altered northern Indiana landscape coupled with southern Indiana's bluffs. As a meeting place for major conventions and trade shows in an urban environment, the building's design captures the welcoming gesture of a handshake.

A new entrance to the Convention Center is defined by a clear glass cube that provides expansive views in and out of the facility, encouraging both conventioneers and local residents to explore the new space. The glass cube is supported by four white steel columns, which can be illuminated with colors inspired

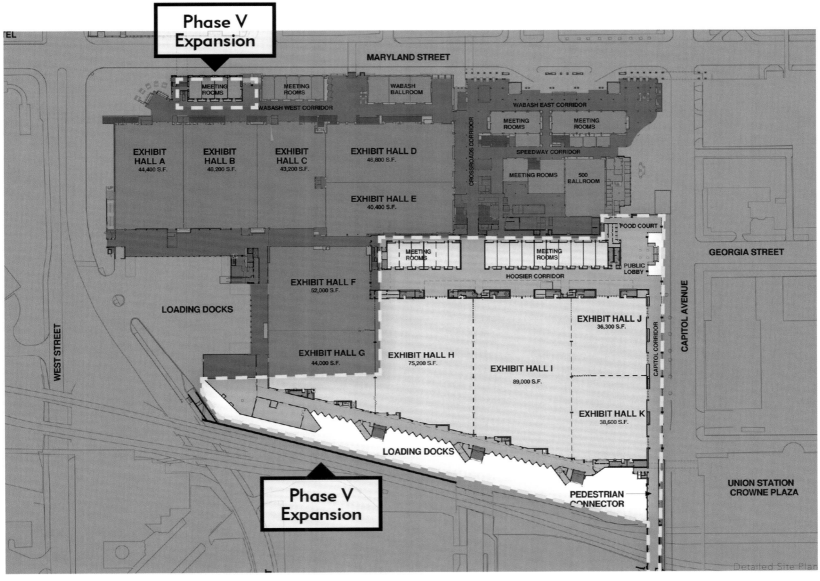

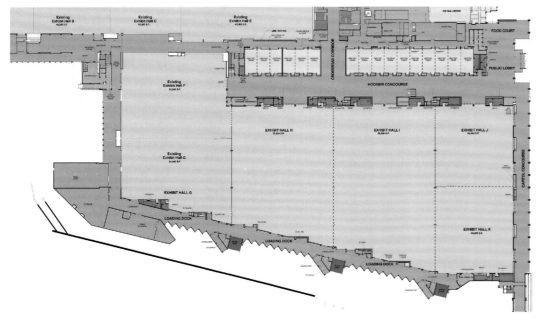

Main Level Floor Plan

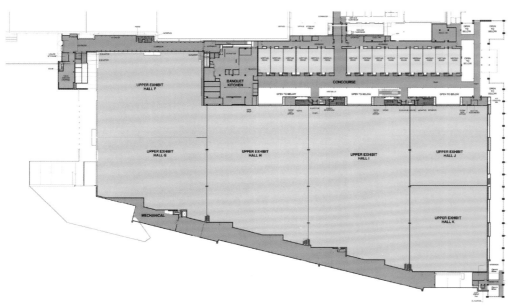

Second Level Floor Plan

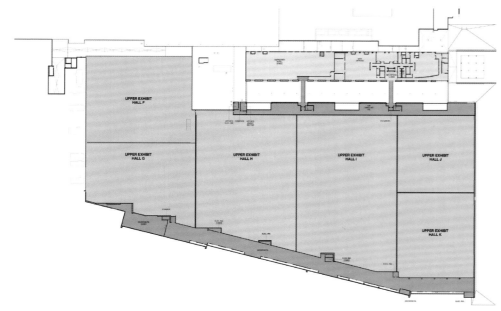

Third Level Floor Plan

by the convention taking place inside. As a meeting place for major conventions and trade shows in a urban environment, the canopy over the cube extends a welcoming gesture to visitors.

The materials used in the expansion integrate the new facility with the existing Center through brick, limestone and curtain wall glazing. The Convention Center's interior design embodies the concept of Indiana as the "Crossroads of America". Interior finishes and the overhead exposed steel structure reference many roads, iron bridges, rivers and train tracks that cross the state through pattern, color and scale. Indiana's four seasons, natural lakes and waterways inspire the facility's warm interior color palette. Hues of cranberry, gold, warm brown tones, forest green, indigo blue and persimmon flow throughout the space, blending the expansion with the existing facility to create a seamless transition.

A main goal was to create a continuity of exhibit space to have one open, flowing and continuous space. To simplify building navigation and create continuity, the Convention Center expansion's color palette is also integrated into the building's informational signage. Inspired by the entrance canopy profile, small canopies located above meeting room entrances, and larger canopies above exhibit hall entries provide additional visual wayfinding cues. Circulation throughout the Convention Center is further celebrated by the inclusion of hanging bridges and a second floor concourse. The glass railing system provides unimpeded views of the facility's activities and emphasizes overall expansion height. Skylights, exterior clerestory and storefront glazing infuse each level with natural light.

The design team used aerial photographs of Indiana's topography as the inspiration for the floor covering designs found throughout the Center's interior. Pattern, scale and color respond to the space's monumental scale. Carpeting located at meeting room entrances imitates the undulating dune sand patterns found in northern Indiana. Connecting corridors are carpeted with a linear, grid-like pattern that references the state's till plains. Carpeting in pre-function areas is reminiscent of the organic shapes found throughout Indiana's southern plains and lowlands. The Indiana Convention Center expansion project's exterior site design was an integral component to its overall design. So that the building could be woven into multiple layers of Indianapolis' downtown urban fabric, the expansion's exterior gathering spaces were designed to accommodate large volumes of pedestrians, significant bus traffic, a multipurpose Cultural Trail, and a future event street located just outside the expansion's glass-cube entry.

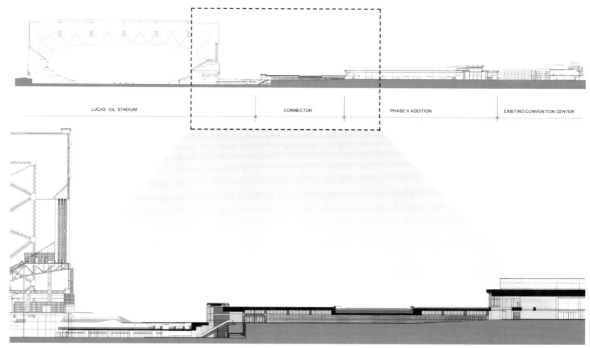

Sections

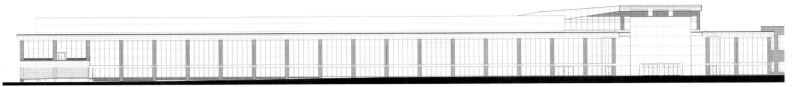
Main Entry Level Elevation

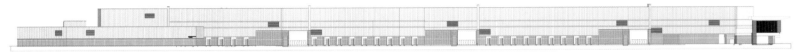
Loading Dock Elevation

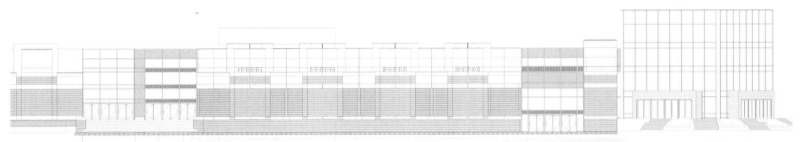
Northwest Addition Elevation

081

Keywords

Simplified Detail
Energy Saving
Specific Materials

The Barnes Foundation

Location: Philadelphia, PA, U.S.A.
Architectual Design: Tod Williams Billie Tsien Architects (TWBTA)
Landscape Design: OLIN
Photography: The Barnes Foundation

In keeping with the solar orientation of the galleries, the rooms will face south along the Benjamin Franklin Parkway, offering a view of the London plane trees along the road. The Collection Gallery has been designed with simplified detail to provide better luminosity for the artwork. Such details as lightening the finish on the wood, simple floor patterns and re-shaping the ceiling to distribute artificial light helped brighten and clarify the viewing within the galleries. The second floor galleries have a clerestory that draws toplight into the spaces and is diffused through louvers.

The new building aims for LEED Platinum certification from the United States Green Building Council (USGBC). The architects employed specific materials and strategies to achieve these results: low or no-VOC products, FSC certified woods, recycled products and reclaimed materials, demolition recycling, energy savings with a 40% reduction in energy use, photovoltaic panels that add up to twelve thousand square feet of the Light Canopy, and landscape irrigation provided by a 40,000 gallon rainwater collector and cistern.

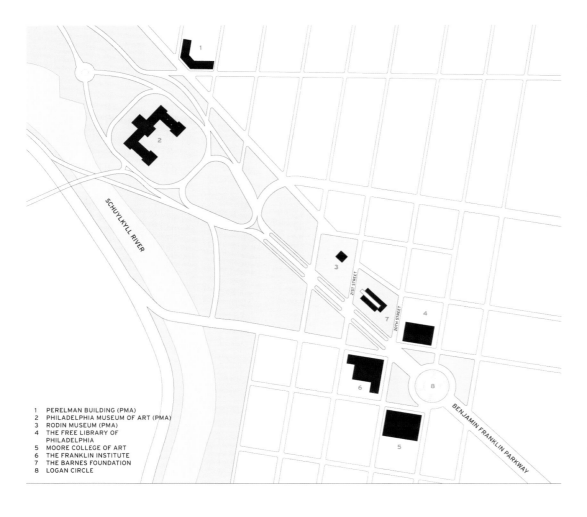

1 PERELMAN BUILDING (PMA)
2 PHILADELPHIA MUSEUM OF ART (PMA)
3 RODIN MUSEUM (PMA)
4 THE FREE LIBRARY OF PHILADELPHIA
5 MOORE COLLEGE OF ART
6 THE FRANKLIN INSTITUTE
7 THE BARNES FOUNDATION
8 LOGAN CIRCLE

The entry experience was designed to create a tranquil environment for visitors. The materials chosen provide a serene environment: transparent glass, Belgian linen and oak wood battens, a walnut staircase, translucent green curtain, sand-blasted architectural concrete and limestone. The lobby provides access to information, tickets, membership, program and event information.

The education spaces are supplemental teaching areas located within the Collection Gallery in the lower level of the building. These spaces include a generous lobby with comfortable seating and a library shelf of Collection books, access to a 150-seat auditorium, two seminar rooms and a coffee bar. The library wraps the Gallery Garden and provides access to the outdoor landscaped spaces. The education area can accommodate discussion, lecture and digital instruction. Similar materials incorporated into the entry are used in the auditorium: limestone, acoustic panels of Belgian linen and white oak and plaster.

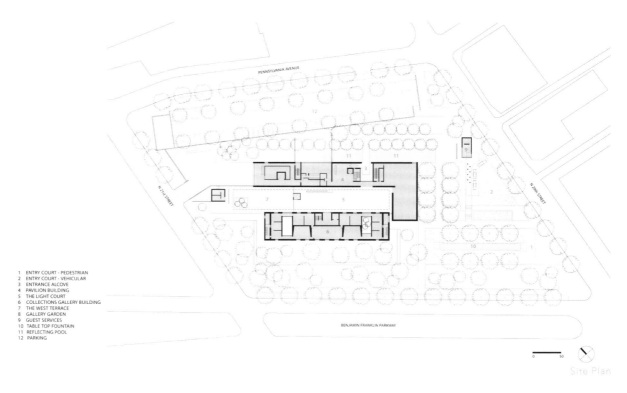

1. ENTRY COURT - PEDESTRIAN
2. ENTRY COURT - VEHICULAR
3. ENTRANCE ALCOVE
4. PAVILION BUILDING
5. THE LIGHT COURT
6. COLLECTIONS GALLERY BUILDING
7. THE WEST TERRACE
8. GALLERY GARDEN
9. GUEST SERVICES
10. TABLE TOP FOUNTAIN
11. REFLECTING POOL
12. PARKING

Site Plan

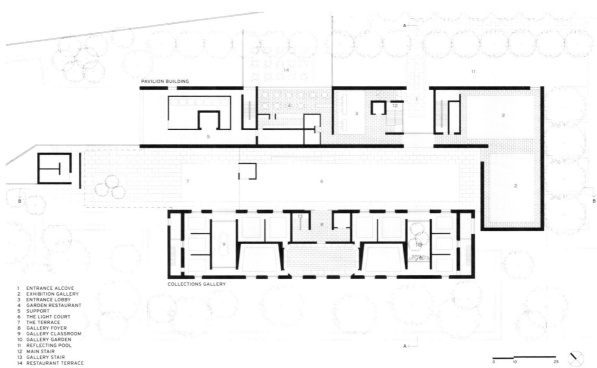

1. ENTRANCE ALCOVE
2. EXHIBITION GALLERY
3. ENTRANCE LOBBY
4. GARDEN RESTAURANT
5. SUPPORT
6. THE LIGHT COURT
7. THE TERRACE
8. GALLERY FOYER
9. GALLERY CLASSROOM
10. GALLERY GARDEN
11. REFLECTING POOL
12. MAIN STAIR
13. GALLERY STAIR
14. RESTAURANT TERRACE

Plan—Floor 1

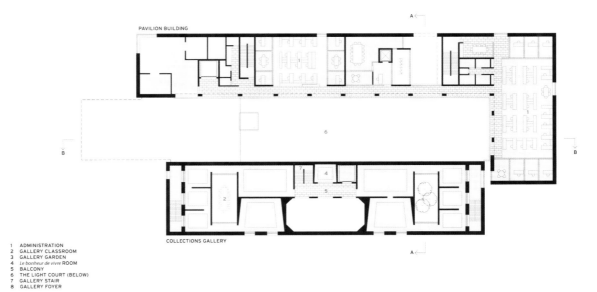

1. ADMINISTRATION
2. GALLERY CLASSROOM
3. GALLERY GARDEN
4. *Le bonheur de vivre* ROOM
5. BALCONY
6. THE LIGHT COURT (BELOW)
7. GALLERY STAIR
8. GALLERY FOYER

Plan—Floor 2

085

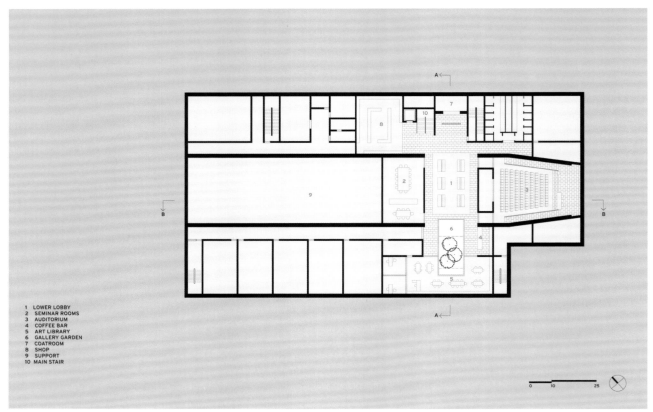

Plan—Floor LL

1 LOWER LOBBY
2 SEMINAR ROOMS
3 AUDITORIUM
4 COFFEE BAR
5 ART LIBRARY
6 GALLERY GARDEN
7 COATROOM
8 SHOP
9 SUPPORT
10 MAIN STAIR

086 | Art Museum

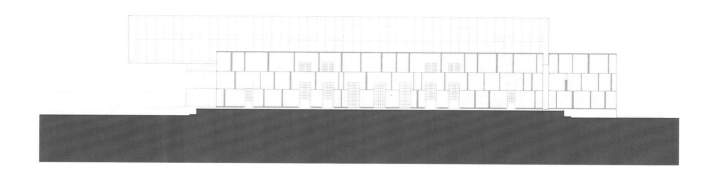

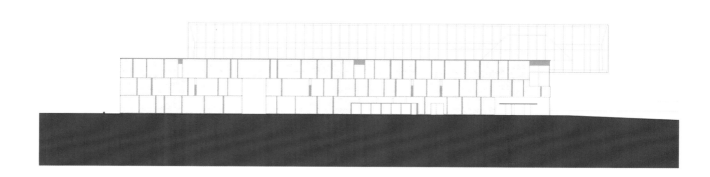

1 ENTRY ALCOVE
2 THE LIGHT COURT
3 THE LIGHT CANOPY
4 GALLERY GARDEN
5 LOWER LOBBY

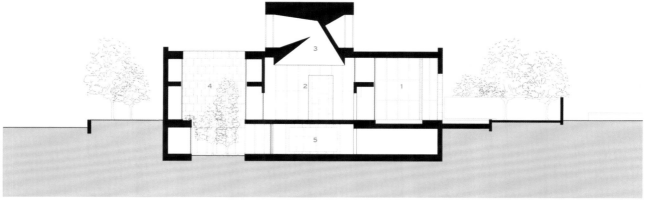

Section A-A: North/South

1 THE LIGHT COURT
2 THE TERRACE
3 THE LIGHT CANOPY
4 ADMINISTRATION
5 EXHIBITION GALLERY
6 AUDITORIUM
7 LOWER LOBBY
8 SEMINAR ROOMS
9 SUPPORT

Section B-B: East/West

Keywords

Sustainability
Resource Saving
Energy Efficiency

Pavilion 4

Location: Shanghai, China
Architectual Design: HMA Architects & Designers
In Collaboration with: Tongji Architectural Design (Group)
Area: 10,200 m²
Photography: Lvfeng photography Studio, Gengtao, Li Xuefeng

The architects had designed a pavilion in Shanghai World EXPO 2010. This building was used as an industrial factory in the 1970s. It is converted from old factory to exhibition center. The EXPO site was situated in center of Shanghai city and was separated both side of Huangpu River. Almost of all national pavilions was designed in Pudong district that was vacant up to 2000s. On the other hand, Puxi district is the site which was used as an industry district. Therefore the architects started to consider how to reuse an old dock factory.

The aim of the pavilion is to exhibit a practical example for sustainable city. The design intention is to reuse existing structure and material as much as possible, and reduce many scrap materials by half. After the architects researched the existing building they found that original wall needs to remove. It was because it couldn't stand for. When all walls were removed, it tried to keep original brick in the site which would be used afterward. Then these original bricks were piled up again with some decorations and patterns. Moreover roof window was added on the top for getting sunlight and natural ventilation and dull flat roof shape was redesigned to random shape.

One another building had removed except structure columns and beams. Then gourd shape of void was inserted into existing structure. That gourd shape promotes gravity ventilation. Both building will be use as an exhibition center after the Shanghai World EXPO was finished.

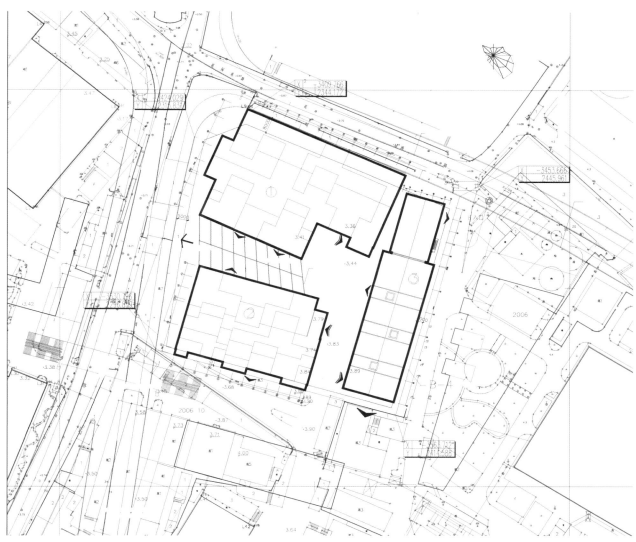

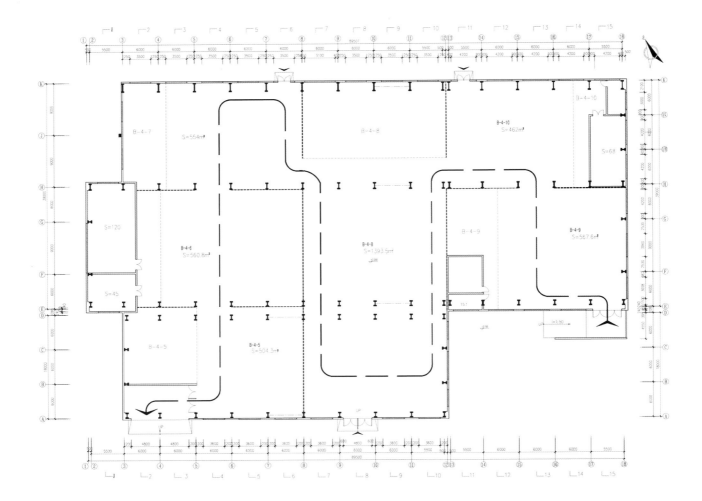

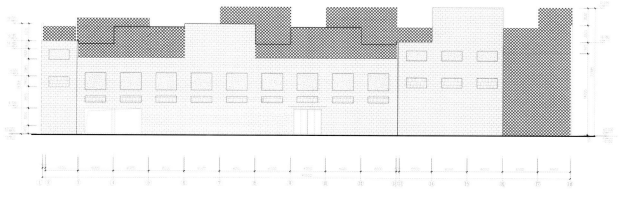
Elevation 1

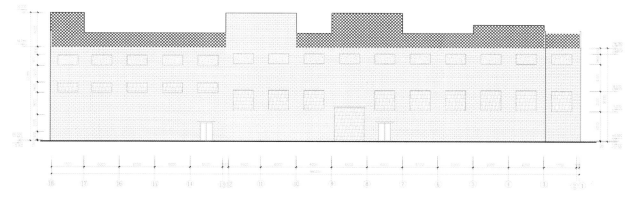
Elevation 2

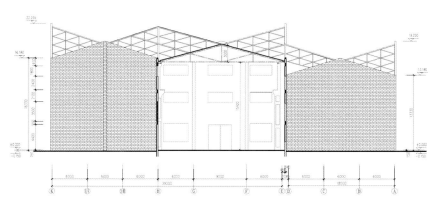
Section 1

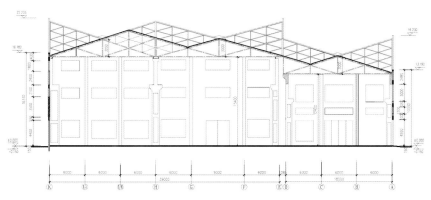
Section 2

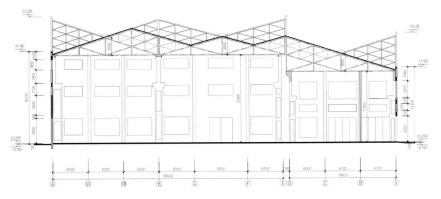
Section 3

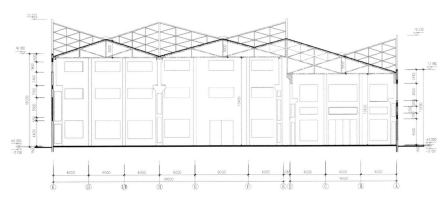
Section 4

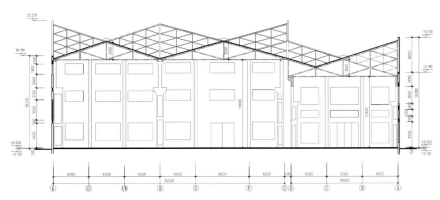
Section 5

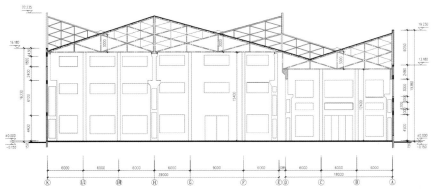
Section 6

103

Cultural Activity Building

Public Use
Cultural Activities
Themed Design

Keywords

Floating Spaces
Multifaceted Volume
Sandwich Structure

Dalian International Conference Centre

Architects: Coop Himmelb(l)au
Client: Dalian Municipal People's Government, P.R. China
Location: Dalian, China
Design Principal: Wolf D. Prix
Project Partner: Paul Kath (until 2010), Wolfgang Reicht
Project Architect: Wolfgang Reicht
Design Architect: Alexander Ott
Design Team: Quirin Krumbholz, Eva Wolf, Victoria Coaloa
Site Area: 40,000 m^2
Gross Floor Area: 117,650 m^2
Photography: Duccio Malagamba

The building has both to reflect the promising modern future of Dalian and its tradition as an important port, trade, industry and tourism city. The formal language of the project combines and merges the rational structure and organization of its modern conference centre typology with the floating spaces of modernist architecture.

The urban design task of the Dalian International Conference Center is to create an instantly recognizable landmark at the terminal point of the future extension of the main city axis. As its focal point the building will be anchored in the mental landscape of the population and the international community.

The footprint of the building on the site is therefore arranged in accordance with the orientation of the two major urban axis which merge in front of the building.

The cantilevering conference spaces that penetrate the facades create a spatially multifaceted building volume and differentiate the close surroundings.

The various theatres and conference spaces are covered by a cone-shaped roof screen. Through controlled daylight input good spatial orientation for the visitors and atmospheric variety is assured.

The project combines the following functions within one hybrid building with synergetic effects of functionality and spatial richness.
· Conference Center
· Theater and Opera House
· Exhibition Center
· Basement with Parking, Delivery and Disposal

A public zone at ground level allows for differentiating accessibility for the different groups of users. The actual performance and conference spaces are situated at +15,30 m above the entrance hall. The grand theater, with a capacity of 1,600 seats and a stage tower, and the directly adjacent flexible conference hall of 2,500 seats, are positioned at the core of the building.

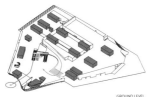 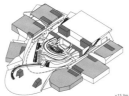
Circulation opera

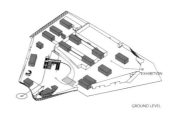 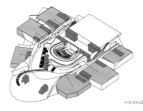 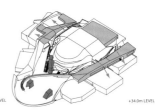
Circulation conference halls

Circulation
Study model 1:200

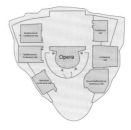

Circulation of opera and conference halls

With this arrangement the main stage can be used for the classical theater auditorium as well as for the flexible multipurpose hall. The main auditorium is additionally equipped with backstage areas like in traditional theaters and opera houses. This scheme is appropriate to broaden the range of options for the use of this space: from convention, musical, theater even up to classical opera, with very little additional investment.

The smaller conference spaces are arranged like pearls around this core, providing very short connections between the different areas, thus saving time while changing between the different units. Most conference rooms and the circulation areas have direct daylight from above.

Through this open and fluid arrangement the theater and conference spaces on the main level establish a kind of urban structure with "squares" and "street spaces". These identifiable "addresses" facilitate user orientation within the building. Thus the informal meeting places, as well as chill-out and catering zones, and in between the halls, gardens with view connection to outside are provided as required for modern conference utilization.

The access to the basement parking garage, truck delivery and waste disposal is located at the southwest side of the site,

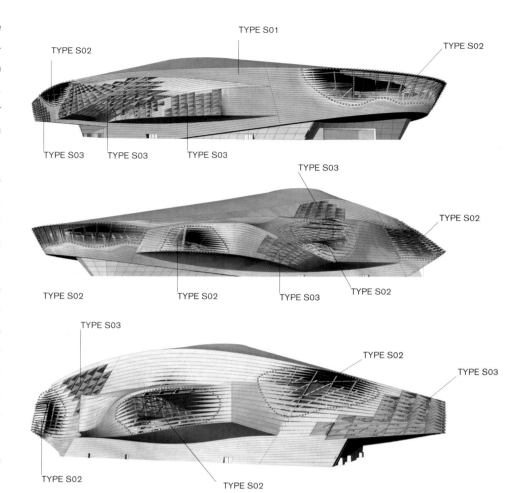

Sea water cooling

Displacement of ventilation systems

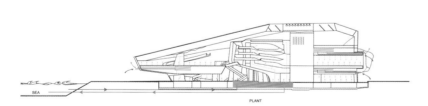

Floor heating/ cooling

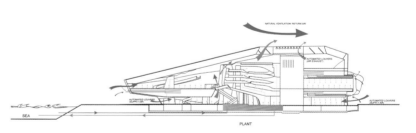

Natural ventilation

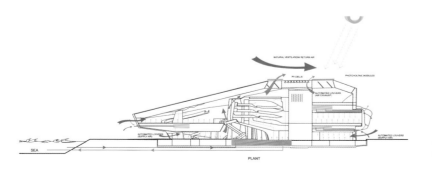

Solar energy/ PV cells

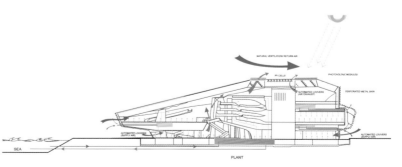

Natural light/ shadering

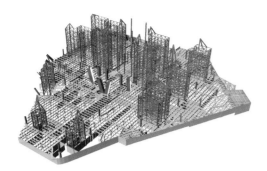
cores and columns
vertical steel concrete bond

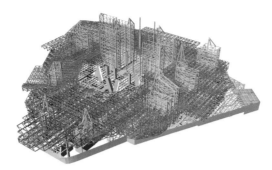
table
spatial steel framework

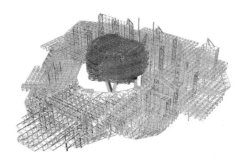
main auditorium
spatial steel structure

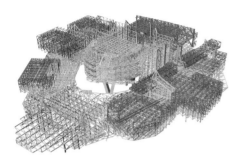
conference boxes
spatial steel structure

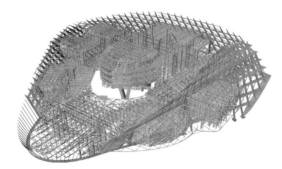
facade
spatial twisted steel framework

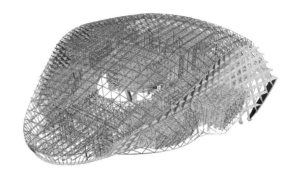
roof structure

thus freeing the front driveway to the entrances from transit traffic. The main entrance from the sea side corresponds to the future developments, including the connection to the future cruise terminal.

Technical, Climatic and Environmental Concept
The focus of the architectural design and project development lies on technology, construction and their interplay. The technical systems fulfil the tasks required for the spatial use of the building automatically, invisibly and silently.

With the Dalian International Dalian Conference Center, these systems work like a hybrid city within a building. For the technical infrastructure of the building this means, that we have to consider a huge amount of people circulating inside the building at the same time, who expect high standards in circulation and comfort as well as a state of the art building with respect to high flexibility, low energy consumption and low use of natural resources.

Technical areas in the basement supply infrastructure within a rectangular grid, mainly inside the vertical cores. In particularly the conference zone has to be provided with a sufficient amount of air in order to maintain a high level of thermal and acoustical comfort. Therefore the conditioned air will be silently injected into the rooms via inflated double flooring underneath the seating. Air blowout units inside the stairs will ensure consistent air distribution. Due to the thermal uplift, the heat of the people ascends to the ceiling and is extracted by suction.

One of the major tasks of sustainable architecture is the minimisation of energy consumption. A fundamental contribution is to avoid considerable fluctuations in demands during the course of the day. Therefore it is essential to integrate

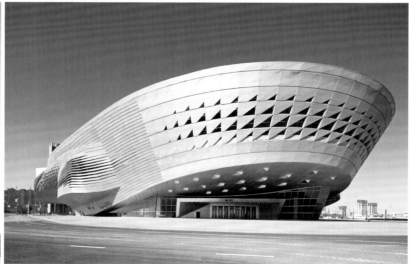

the natural resources of the environment like:

· Use the thermal energy of seawater with heat pumps for cooling in summer and heating in winter.

· General use of low temperature systems for heating in combination with activation of the concrete core as thermal mass in order to keep the building on constant temperature.

· Natural ventilation of the huge air volumes within the building allows for minimization of the mechanical apparatus for ventilation heating and cooling. The atrium is conceived as a solar heated, naturally ventilated sub climatic area.

· In the large volume individual areas can be treated separately by additional measures such as displacement ventilation

· A high degree of daylight use is aspired both for its positive psychological effect and for minimizing the power consumption for artificial lighting

· Energy production with solar energy panels integrated into the shape of the building.

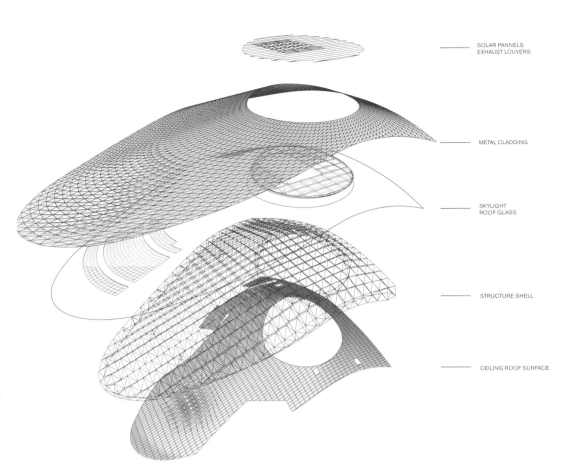

Roof Structure

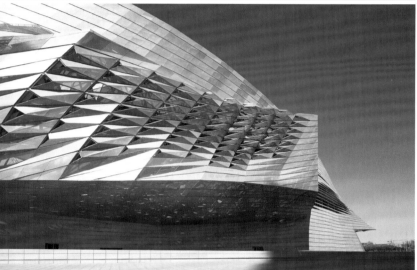

Structural Concept

The structural concept is based on a sandwich structure composed of 2 elements: the "table" and the roof.

Both elements are steel space frames with depths ranging between 5 and 8 meters.

The whole structure is elevated 7 meters above ground level and is supported by 14 vertical composite steel and concrete cores.

A doubly ruled façade structure connects the two layers of table and roof, creating a load-bearing shell structure.

The application of new design and simulation techniques, the knowledge of local shipbuilders to bend massive steel plates, and the consumption of more than 40,000 tons of steel enables breathtaking spans of over 85 meters and cantilevering of over 40 meters.

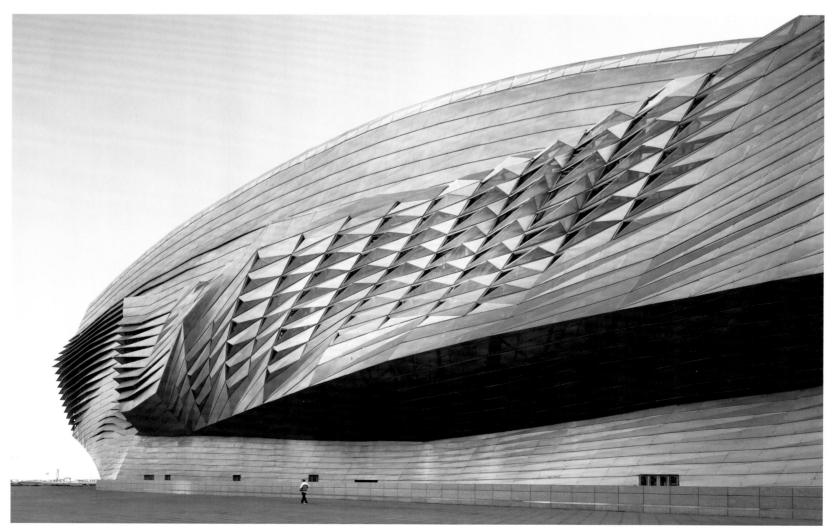

Keywords

Building Volumes
Spatial Experience
Economic & Ecological

BRG Neusiedl am See

Architects: Solid Architecture
Location: Neusiedl am See, Austria
Site Area: 26,270 m^2
Gross Area: 9,206 m^2
Useable Surface: 6,812 m^2
Built Up Area: 5,648 m^2
Building Volume: 41.060 m^3
Photography: Kurt Kuball

Initial Position

The Bundesrealgymnasium (secondary school) in Neusiedl am See is a strictly orthogonal school complex dating from 1972. Three two-storey classroom wings running north-south and a gymnasium wing are connected by a single-storey volume running east-west.

The commission involved refurbishing the school building and adding a new circulation wing to it. The single-phase competition open to entries from throughout the EU which was organized by the Bundesimmobiliengesellschaft (BIG, a quasi-governmental company that manages Austrian publicly owned real estate) was won by the working partnership of SOLID architecture / K2architektur.

Building Volumes and Program of Spaces
The design for the extension preserves the orthogonal character of the existing school.

In place of the single-storey connecting element a new two-storey volume is erected that connects the three classroom wings with each other on both levels. In addition to the central staircase this new building also contains an elevator that provides barrier-free access to the entire school, a multi-purpose hall, classrooms and areas used during breaks from lessons. The latter areas are on both ground and first floor. Large areas of glazing towards the south offer a view of the recess yard planted with large plane trees. A projecting roof slab prevents the rooms from overheating in summer.

A single-storey connecting building running in an east-west direction houses the afternoon care facilities, for which a new internal courtyard was created. The common room for afternoon care can be connected to the dining room and has a terrace in front that faces south, onto the new courtyard.

Spatial Experience – Clear Articulation

The open, transparent connecting wing with the areas for school breaks augments the existing compact – and rather introverted – classroom wings. The extrovert central section that is oriented towards the open spaces increases the overall clarity of the school complex. The publicly used areas and recess areas relate directly to the outdoor spaces.

Colour Concept

The colour concept for the building responds to the impressive existing trees on the site. Borrowing from the shades of the large plane trees in the recess yard, the dominant colours are green, beige and white.

Site Plan

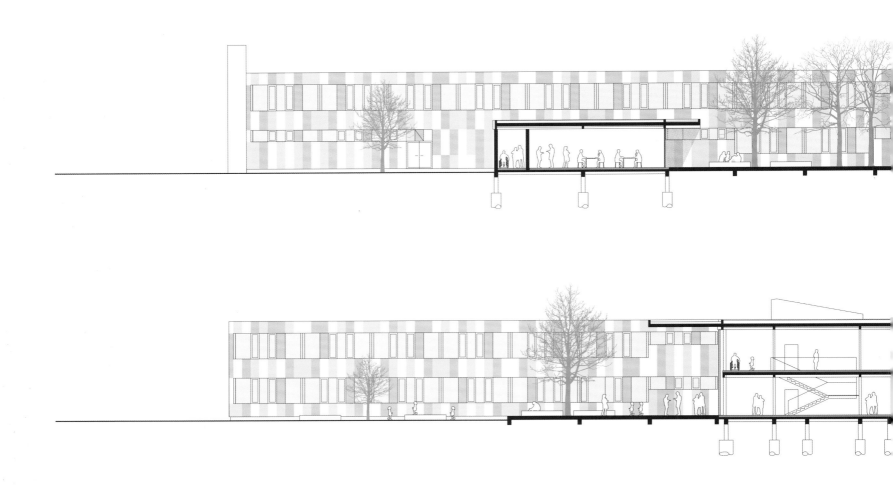

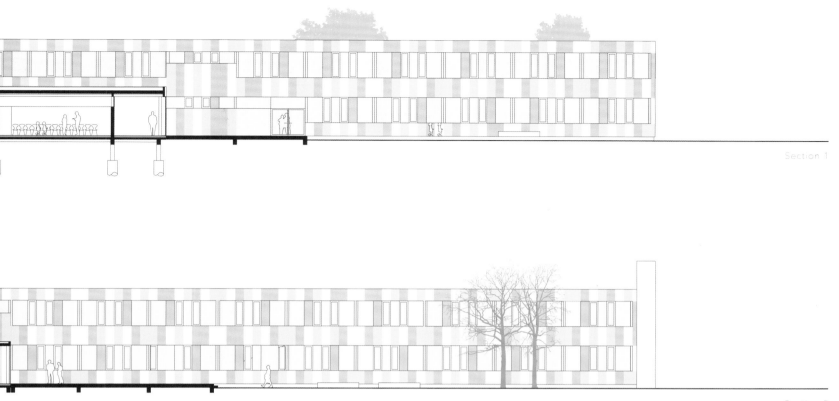

Section 1

Section 2

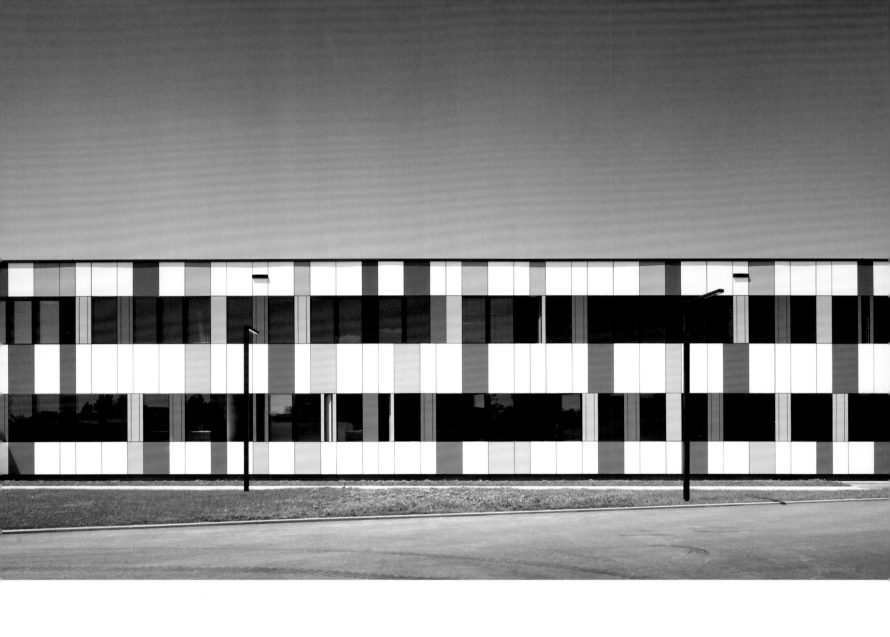

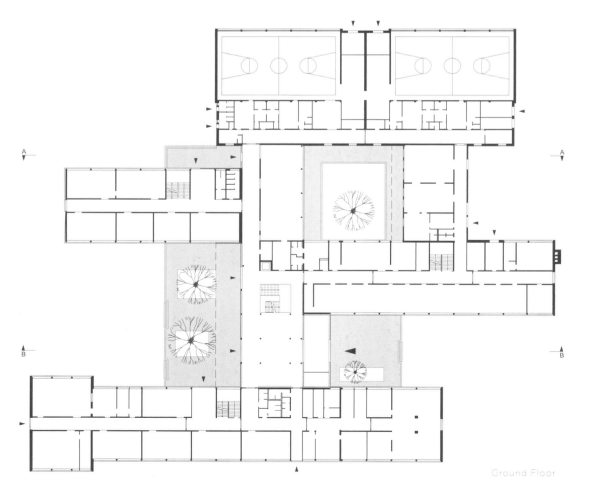

Refurbishment of the Existing Buildings
Particular emphasis is placed on achieving a good energy balance for the building.

The building envelope is upgraded to meet current demands in terms of building physics. The existing classroom wings are given a back-ventilated façade composed of green and white panels.

The lively colour of the design provides an identity for the school as a whole.

In the interior all the floor, wall and ceiling surfaces are renovated and designed in accordance with the colour concept for the school. The façade areas of the gym hall are thermally upgraded by applying a white external insulation and finishing system.

Design of the Outdoor Areas

The design of the outdoor spaces is determined by a large "deck" that adjoins all the public areas on the ground floor and blurs the boundary between indoor and outdoor space. This deck connects inside and outside without a change of level and has the same finish as the areas used for breaks in the interior of the building. Projecting roofs on the south side further blur the boundary between inside and outside.

Economic and Ecological Aspects

The goal of reducing energy demand as far as possible and the aim to use an ecological construction method for the new buildings are both met by using timber composite constructions: the two-storey connecting wing is built of timber and steel, the afternoon care facilities of timber and reinforced concrete. Compared to massive construction this method is considerably more sustainable in ecological terms, especially regarding global warming potential.

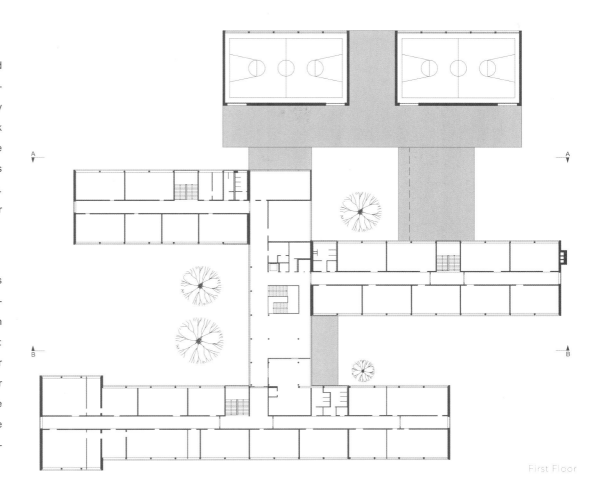

First Floor

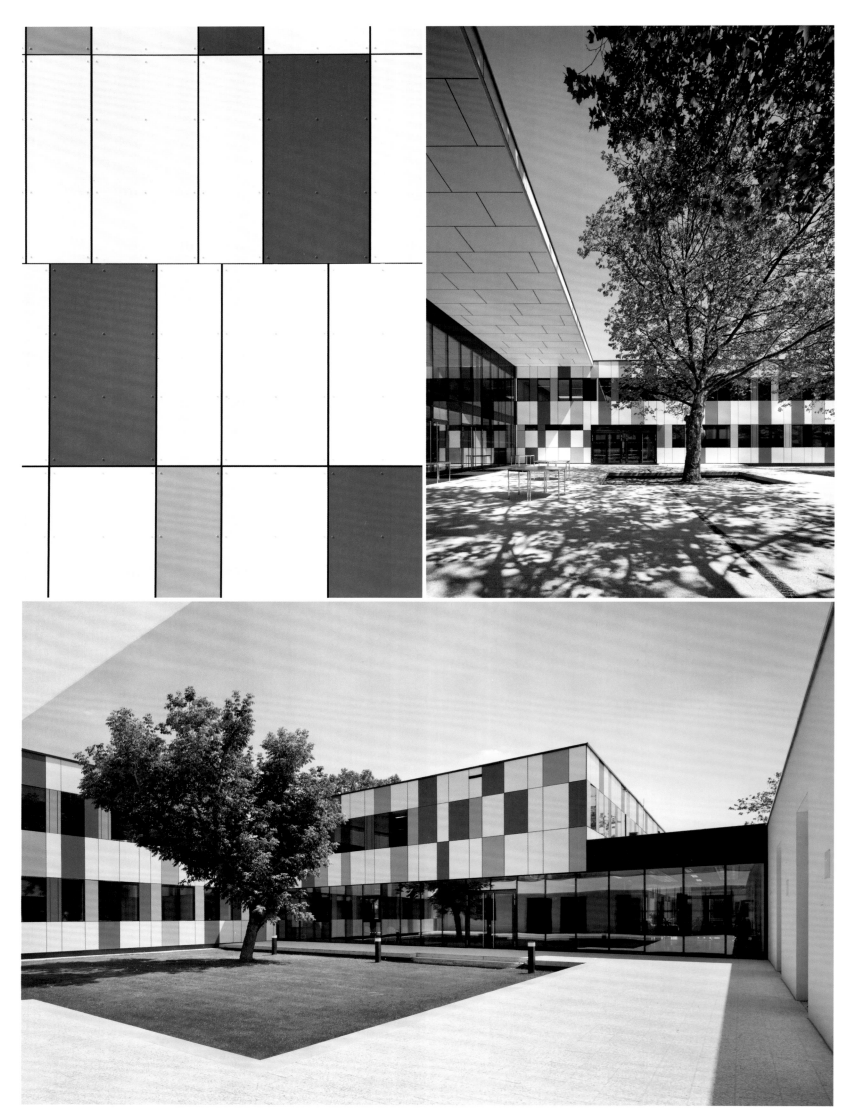

Keywords

Diagonal Views
Prismatic Shape
Glass Facades

Universidad Politecnica de Valencia Expansion

Location: Valencia, Spain
Architects: Corell Monfort Palacios Arquitectos - Vicente Corell Farinós,
            Joaquín Monfort Salvador, José Vicente Palacio Espasa
Developer: Universidad Politécnica de Valencia
Total Area: 17,156.39 m^2
Photography: Mariela Apollonio

The building, located within the grounds of the "Campus de Vera" of the UPV with its main facade facing towards the boulevard "Avenida de los Naranjos", groups in a single volume as the expansion of the ETSIT, the Department of Languages and Language Center of the UPV.

Due to the extensive program that was demanded, the occupation of the total facade fronts forced. The building is divided in two parallel blocks of classrooms and offices linked to the north and south facades, and a central space occupied by circulations, openings and terraced patios which facilitate the entrance of light and generate diagonal views.

A strict modulation, based on the dimensions of the basic cell type that defines the office and orders the whole structure of the building, regulating also the composition and construction of facades

Their prismatic shape is outlined by a continuous surface of concrete that frames the facades that face north-south orientations and emphasizes the separation between the two different institutions in the building with a vertical opening in the facade. A second layer of concrete generates the open space at ground floor that provides access to both centers and connects the square with the internal net of the University Campus.

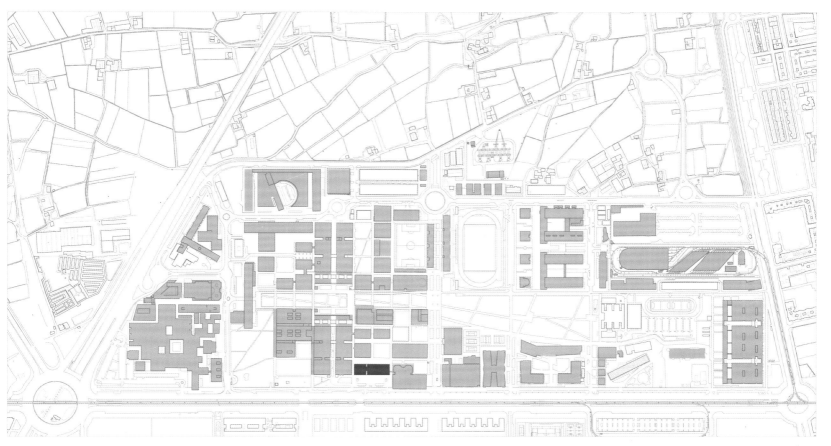

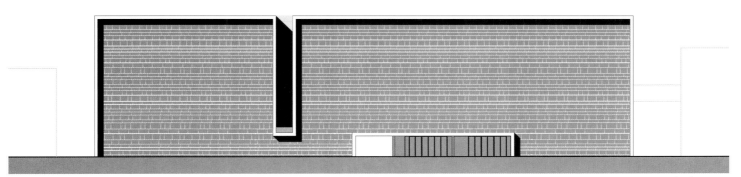

South Elevation

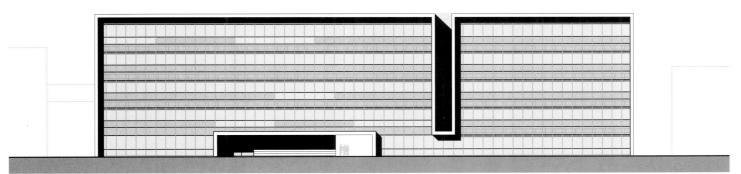

North Elevation

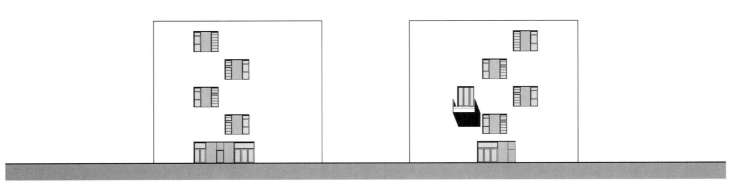

West Elevation · East Elevation

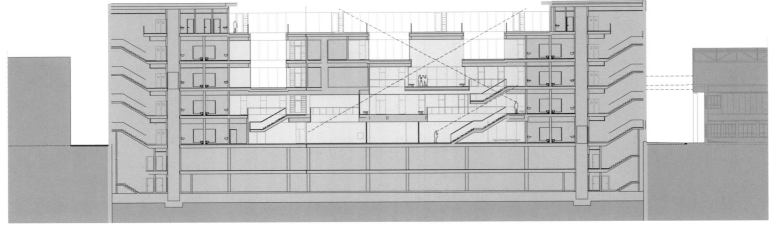
Longitudinal Section

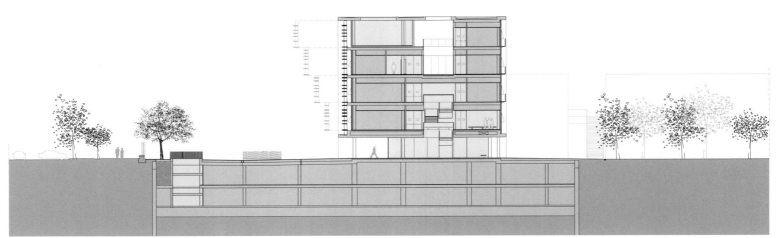
Cross Section

The solution of the north and south facades, meets the requirements imposed by its antagonistic orientations and the willingness to give a unified and clear image. While the north is solved by a continuous glass wall, divided by metal maintenance walkways, the south, also glass facade, is protected by special pieces of precast concrete that act as solar and climatic filter and prevent the vision of the technical utilities that occupy the upper floor.

The white concrete, material that characterizes and gives unity to the image of the building, is used in two different ways: executed in situ with wooden shuttering on the walls and slabs that make up the wrapping surface, and precast for the different concrete pieces that build up the south facade.

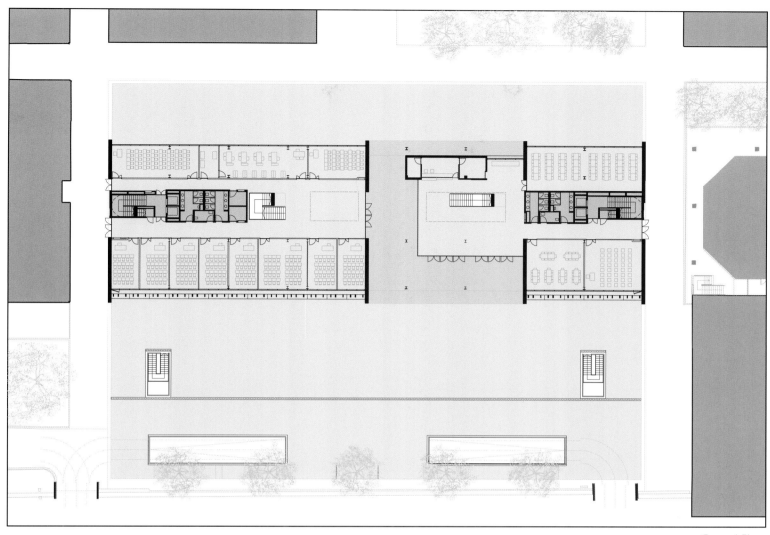
Ground Floor

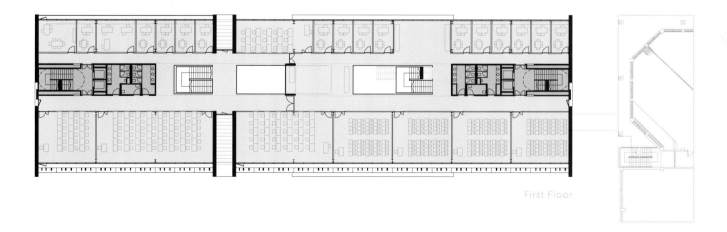
First Floor

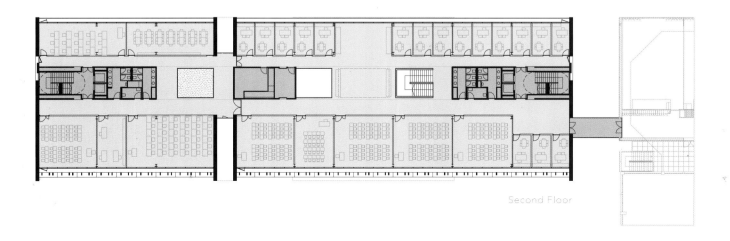
Second Floor

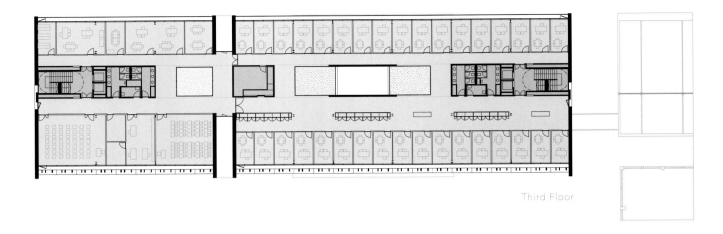

Third Floor

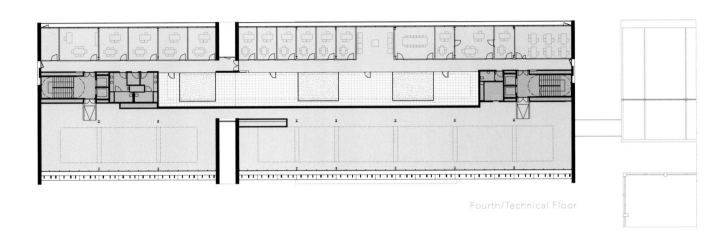

Fourth/Technical Floor

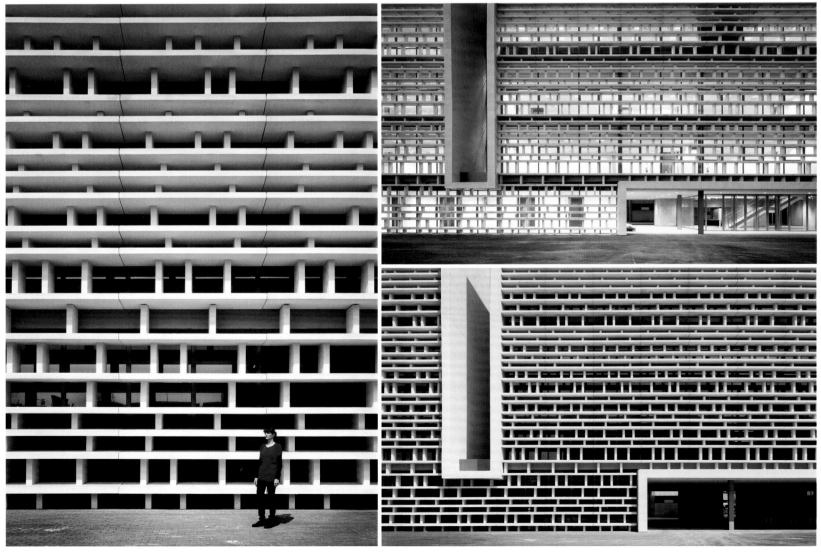

136 | Art Museum

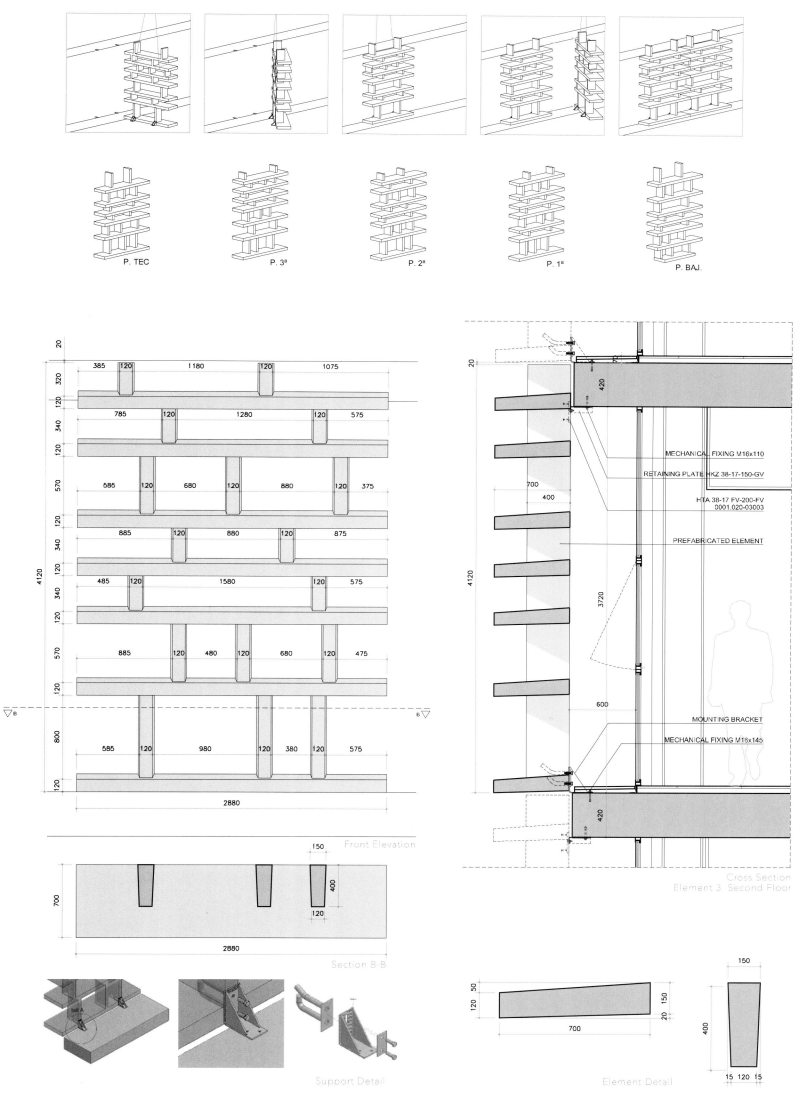

Keywords

Comba Tentes Educational Center
Waterproof Barrier
Continuity

Comba Tentes Educational Center

Location: Ovar, Portugal
Client: Câmara Municipal de Ovar
Architectural Design: CANNATÀ & FERNANDES architects
Site Area: 2, 440.97 m²
Floor Area: 5, 271 m²
Photography: Luis Ferreira Alves, Dario Cannatà, Pepe Barbiere

The project, while maintaining the relations of the preexisting building with the street, adds a new building, formally characterized by its curvilinear geometry which is able to create conditions for the development of the school program while rebalancing all the terrain lines and angles. The different functions are articulated in two buildings (old and new), characterized by the different architectural objectives (rehabilitation and new construction) and the different school program requirements to receive children from the age of three to ten years old.

On the preexisting building, the project recovers the interior court and all the elements that characterize the four facades. The interior spaces are rehabilitated, depending on the functional requirements that the building has to meet, ensuring the balance with their specific architectural character. The functions related to public use were installed in this building, ensuring a smooth relationship between the school and the community - office, meeting room of the parent's association, staff room, office and service areas, library, computer room, curriculum enrichment / music and related services.

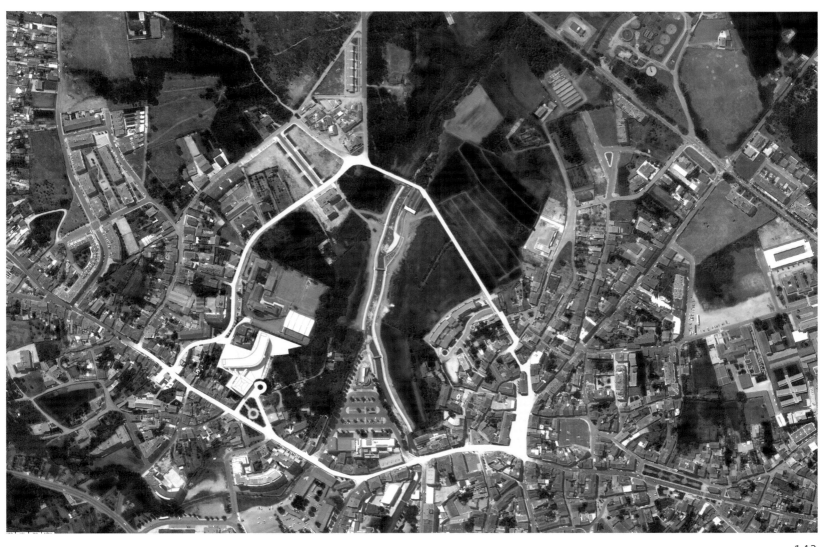

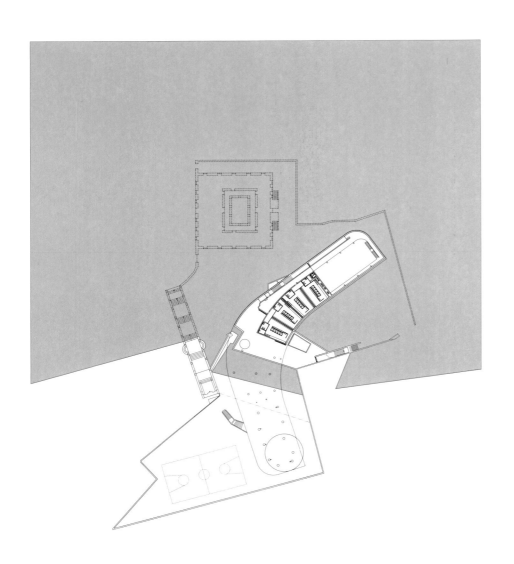

The new building contains the classrooms, gymnasium, kindergarten, canteen, kitchen and related services and multipurpose rooms. This is characterized in the Southwest by a facade comprising a blade system which controls the light and heat, while producing a variation in the building associated with games, creativity and rigor. The Northeast is characterized by a curved facade made of concrete, isolated and covered with tiles. The use of tiles creates a waterproof barrier and establishes a relationship of continuity with an ancient technique well dominated by the constructive culture of Ovar. To ensure continuity of the spaces two covered passages were planned between the buildings in both levels.

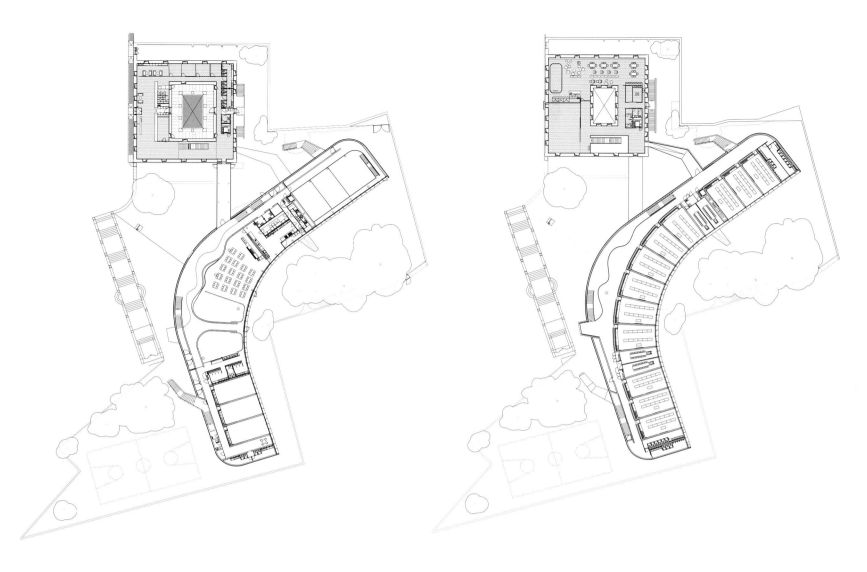

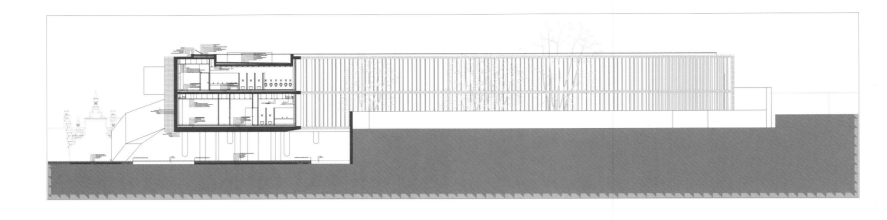
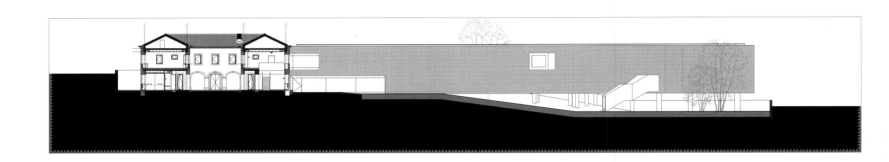

146 | Art Museum

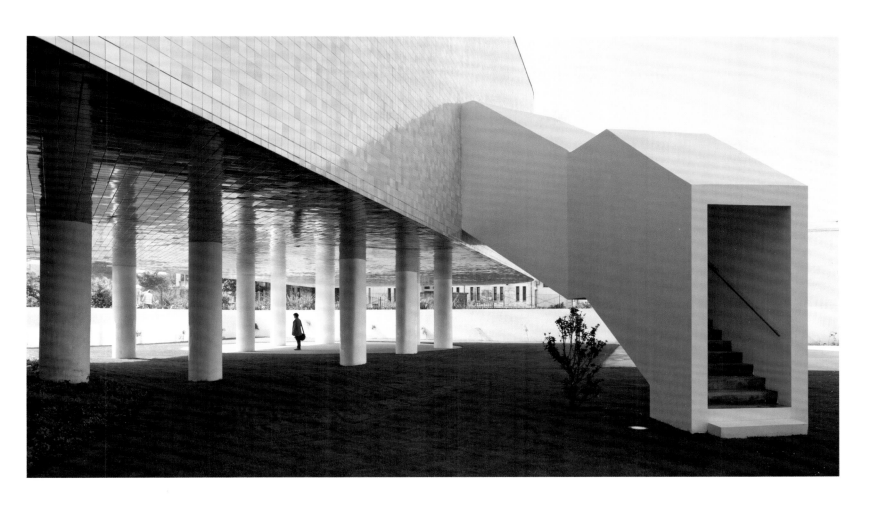

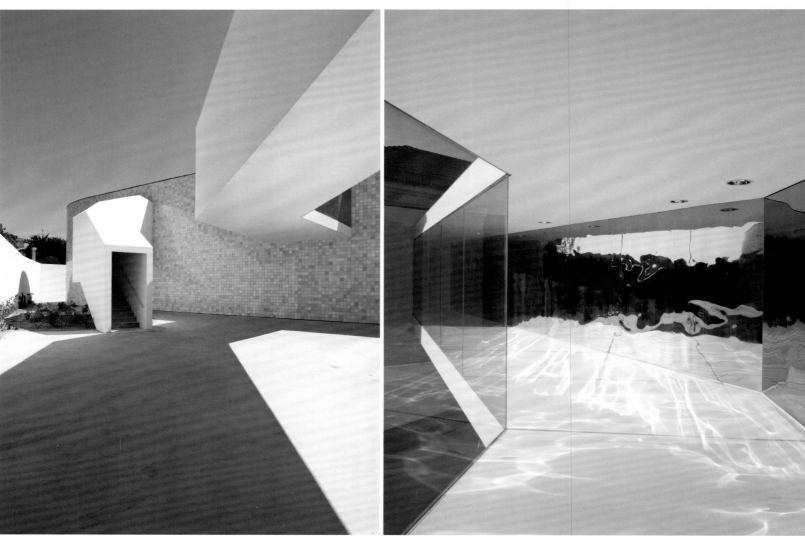

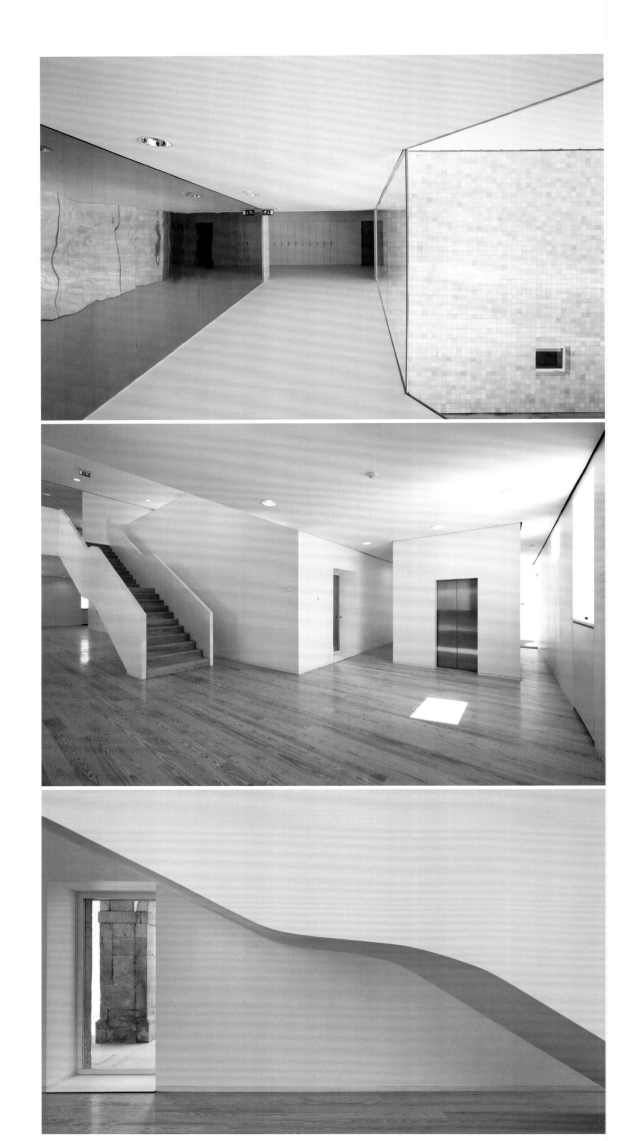

152 | Art Museum

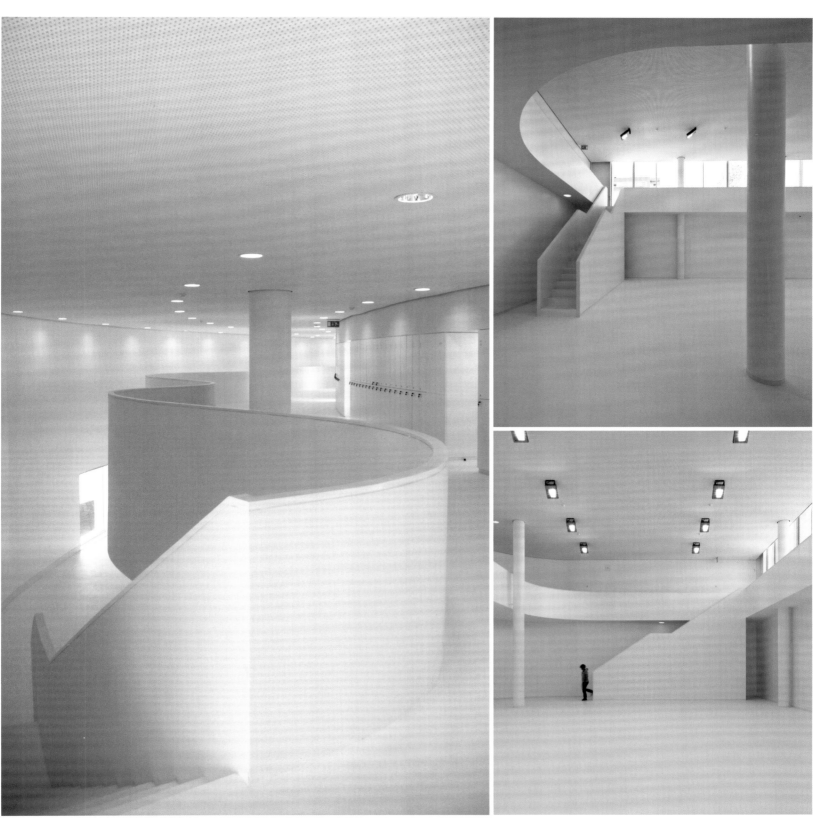

Keywords

Educational Center
Worship Center
Connection Point

Southland Christian Church

Location: Lexington, Kentucky, USA
Architectual Design: EOP Architects
Area: 17,558m^2
Completion: 2013
Photography: Phebus Photography

The LED-lighted sun screen that drapes the exterior of the educational center symbolizes a contemporary stained glass window and the slanted colonnade composition that defines the worship center's exterior is a modern interpretation of the crucifixion. Additional design metaphors include references to the heavens through a pattern of recessed light niches and angular wall planes representing mountains.

The 13,000 m^2 space that formerly housed a large regional department store has been renovated into educational spaces. There are several teaching and group gathering rooms, a nursery, separate learning spaces for pre-school, K-2, grades 3-5 and a large center for student ministries for middle and high school students.

Adjacent to the educational spaces, a 4550 m^2 addition houses the 2,800-seat worship center with state-of-the-art audiovisual technology. Three giant screens simulcast sermons from Southland's main campus while a live band performs on its large stage. An asymmetrical baptistery with a dry area for family viewing mirrors asymmetrical exterior entrances. The curved balcony is fronted by a concrete, cantered facade which helps collect low bass notes to enhance sound quality.

The educational spaces and worship hall are joined by Connection Point, marked by a large, asymmetrical polygon with a nautical symbolism for easy wayfinding. At Connection Point, there is a large café for pre- and post-functions as well as an information desk where visitors' questions are answered.

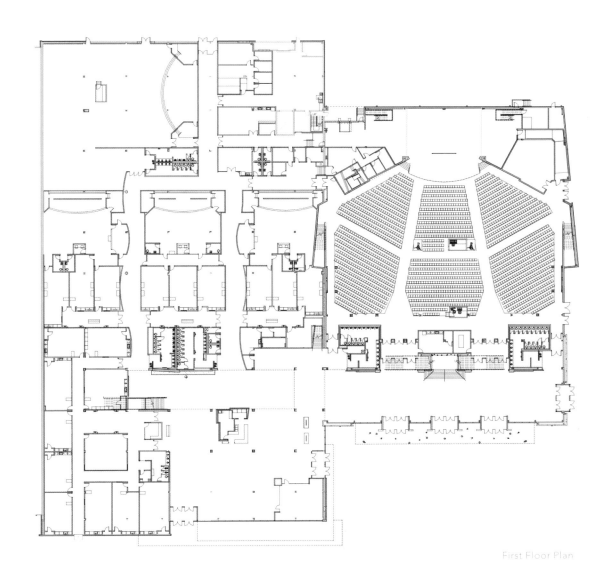

First Floor Plan

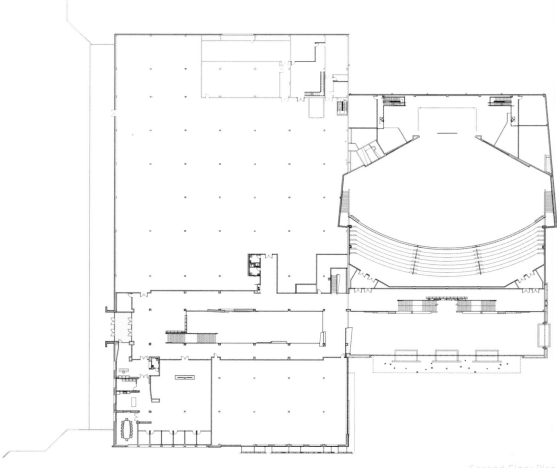

Second Floor Plan

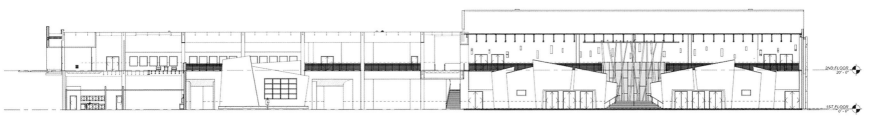
Section

Keywords

Regular Shape
Transparent Facade
Acoustic Barrier

Thebarton Community Centre

Location: South Road, Thebarton, Adelaide, Australia
Client: City of West Torrens
Architectual Design: MPH Architects
Area: 2,898 m²
Photography: David Sievers

The Thebarton Community Centre in Kings Reserve was designed as an iconic "pavilion in the park" in response to the City of West Torren's objectives that is a focus for the community, and a landmark as the northern gateway to the council area. The dominant built form with folded roof planes references movement, flight and the City of West Torrens as a transport hub.

Geometry of the built form has been generated by the irregular site constrained by a major transport corridor to the east and reclaimed pug hole to the west. The parallelogram grid generates a dynamic "tension" with the regular shaped public spaces and contributes to the sense of movement.

The building is arranged with public spaces orientated to the park and support areas providing a buffer to the traffic noise along the South Road boundary.

The transparent facade provides the building with an animated character exhibiting the ever changing internal activities.

Glazing and shading elements were designed based on thermal and solar modelling to achieve a balance between views to the park and minimizing thermal transmission.

The recycled brick facades references the old brickworks site, provides an acoustic barrier to South Road and screens services areas.

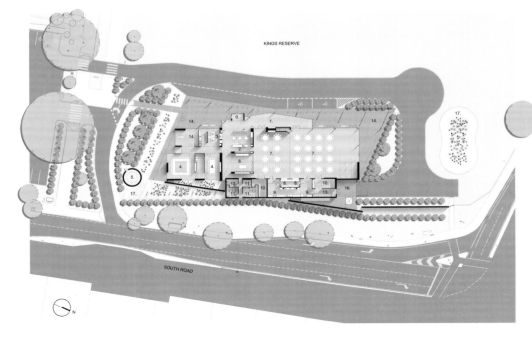

0 Entry
1 Foyer
2 Kitchenette
3 Administration
4 Meeting Room
5 Multipurpose Room
6 Green Room
7 Meeting Hall
8 Bar
9 Kitchen
10 Store
11 Toilet
12 Cleaner's Room
13 Delivery
14 Verandah
15 Rain Water Tank
16 Service Yard
17 Rain Garden

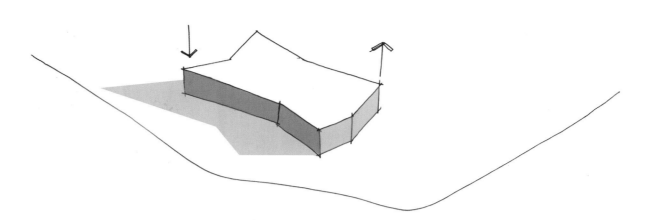

The building turn its back to the traffic lifting the roof to the parkland to engage with its surroundings.

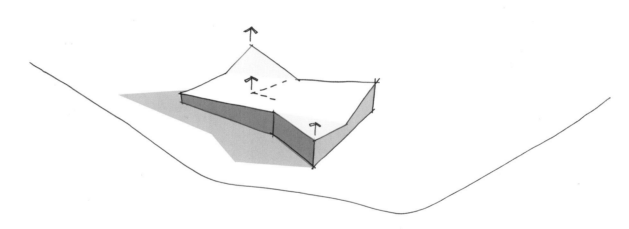

Lifting in the roof winglets maximises the daylight

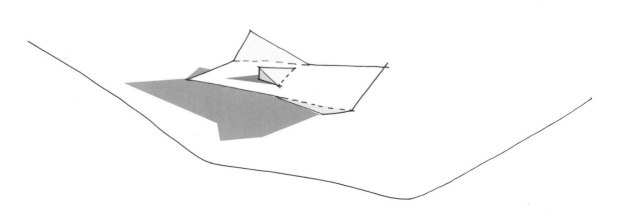

Folded roof planes and wide overhangs reference movement, flight and the City of West Torrens as a transport hub in Adelaide.

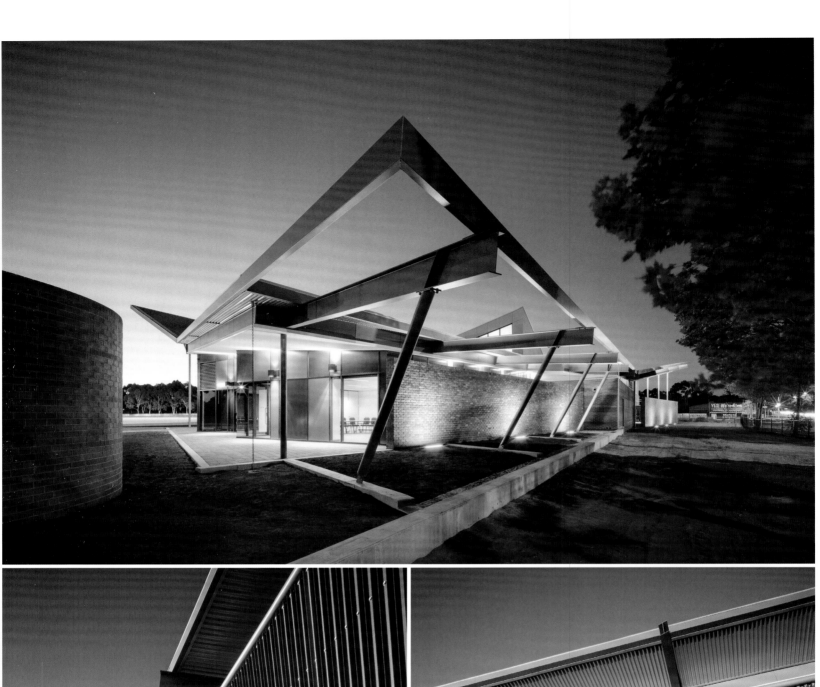

164 | Art Museum

Keywords

Contextual Integration
Duality
Architectural Statement

Le Temps Machine

Location: Joué-lès-Tours
Client: TOUR(S) PLUS (Tours Conurbation Committee)
Architectual Design: Moussafir Architectes Associés
Design Team: Jacques Moussafir avec Nicolas Hugoo, Alexis Duquennoy, Narumi Kang,
      Sofie Reynaert, Jérôme Hervé and Virginie Prié
Area: 1,753 m^2
Photography: Hervé Abbadie, Jérôme Ricolleau

Le Temps Machine (The Time Machine) is a contemporary music venue replacing the 1950s Youth Centre in the French town of Joué-Lès-Tours, an opaque, introverted building that failed to interact with its surroundings and no longer met present-day standards and requirements. Moussafir Architectes suggested maintaining the structure's most salient features, such as the prow-like shape of the auditorium. Today it houses a 150-strong cabaret-style space (the Club), while the Grand Hall for a standing audience of 650 is placed behind it, with rehearsal rooms tucked right below the Grand Hall.

"Every time we design a concert venue, we try to open it towards the surrounding context." says Jacques Moussafir. For stronger contextual integration, the project divides into two distinctive parts. Largely transparent, the horizontal base in concrete and glass contains a fluid, open interior space; deep projecting eaves create a welcoming feel. By contrast, the roof reveals three opaque, powerfully articulated volumes – the Club, the Grand Hall, and the resource centre – that seem to burst through their flat pedestal. "The contradictory image we were aiming at is that of a unique yet familiar object, the one that challenges yet invites appropriation." comments the architect.

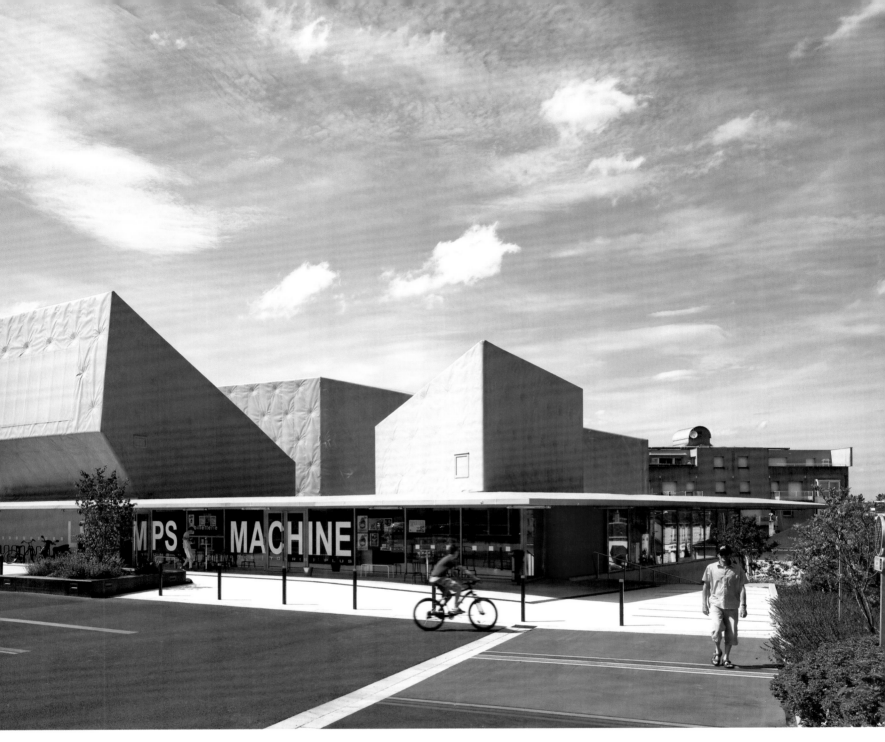

True to his taste for paradox, in Le Temps Machine Moussafir inverts conventional exterior/interior logic. The building appears hard inside and soft outside. The interior is defined by raw concrete, glass and stainless steel. (The thermal inertia of 40 cm-thick concrete walls efficiently reduces the need for artificial cooling, which is fundamental for a concert hall.) The red-and-black color scheme of performance halls is juxtaposed to the total whiteness of the roof, entirely covered with an FPO membrane. The membrane is stretched over exterior insulation and fixed like upholstery: besides enhancing the "soft feel", this not only proved the best way to deal with the material originally intended for horizontal surfaces, but also used here for a bold experiment with three-dimensional volumes.

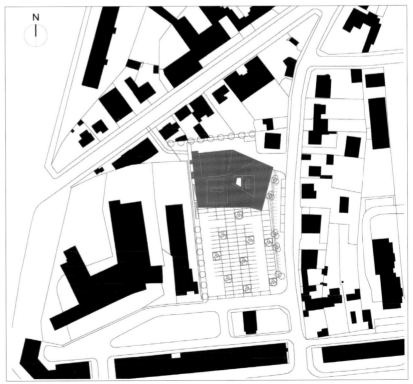

Site Plan

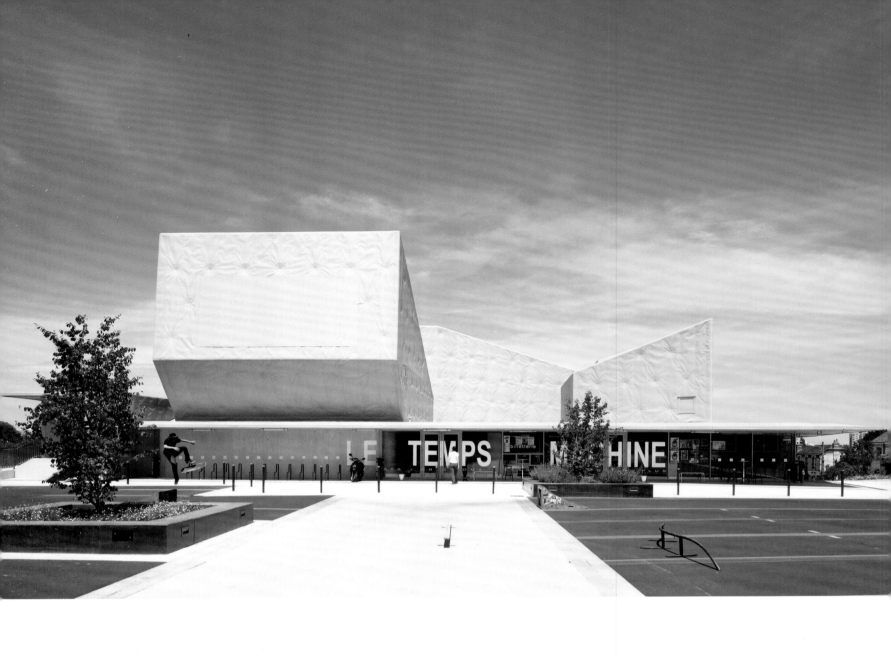

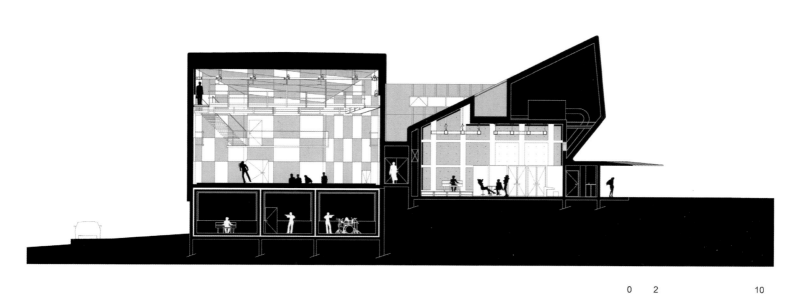
Section 1: the Grand Hall with rehearsal studios below, and the Club

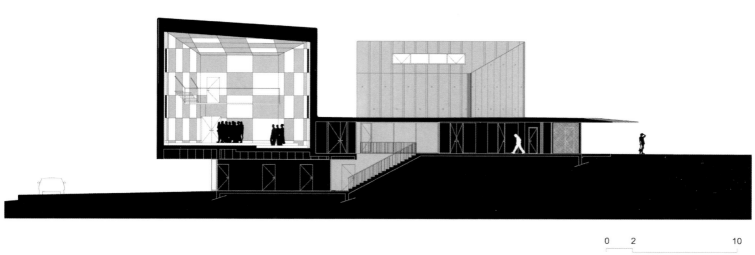
Section 2: the Grand Hall and the patio

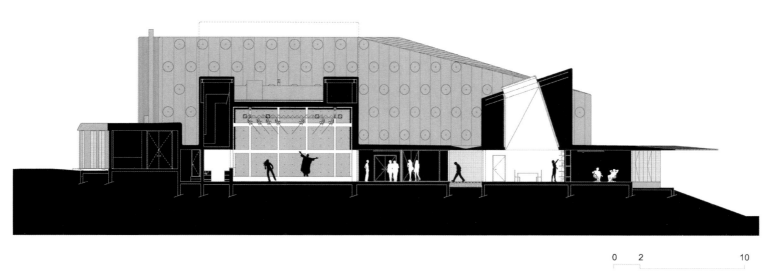
Section 3: the Club and the resources center

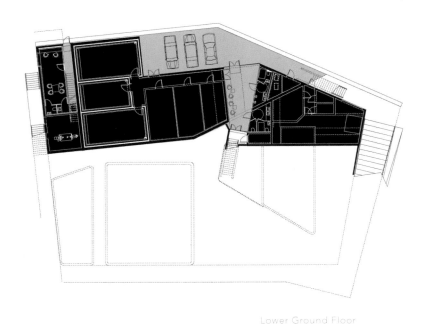
Lower Ground Floor

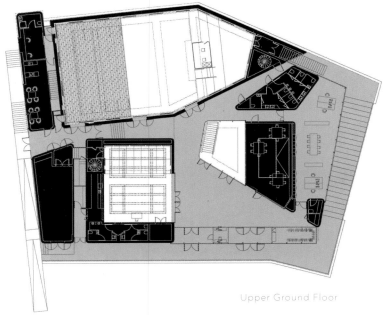
Upper Ground Floor

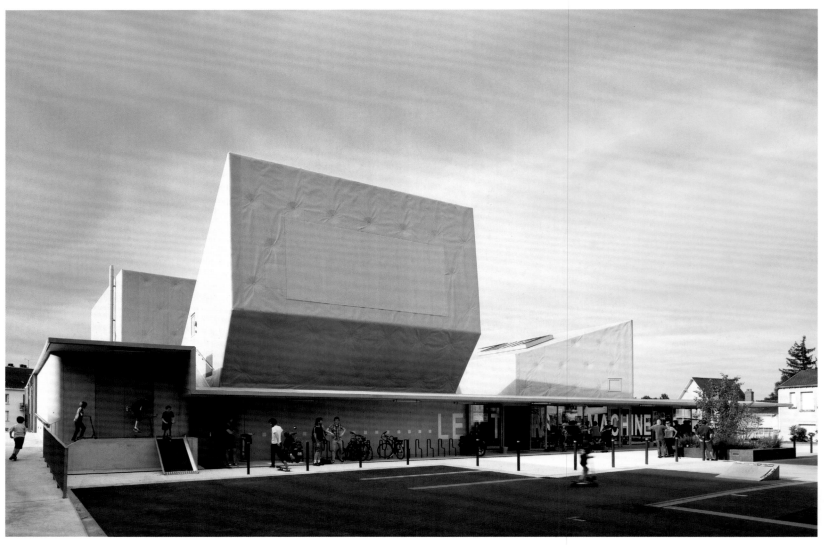

170 | Art Museum

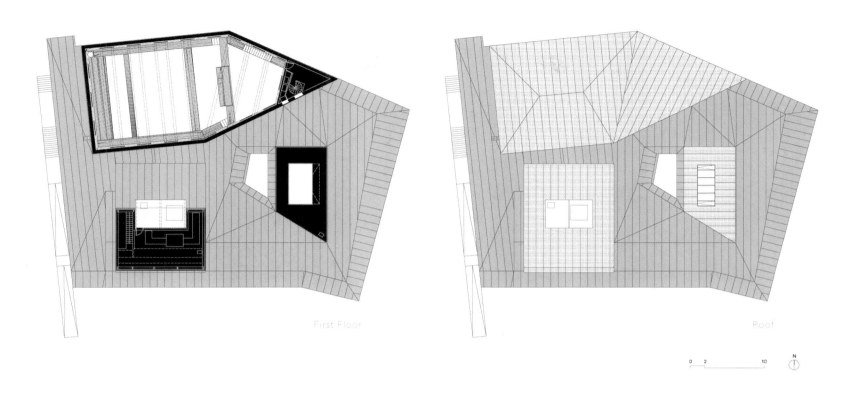

First Floor · Roof

171

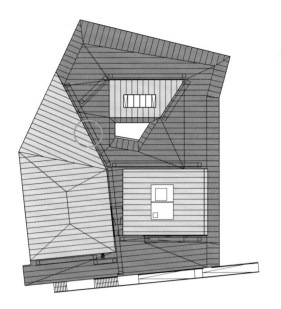

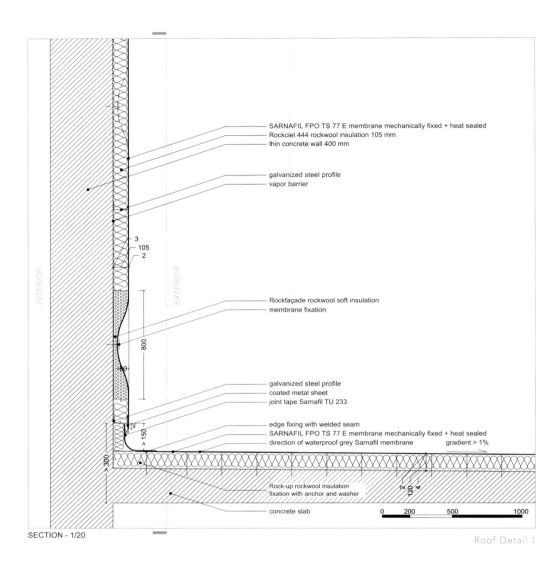

SARNAFIL FPO TS 77 E membrane mechanically fixed + heat sealed
Rockciel 444 rockwool insulation 105 mm
thin concrete wall 400 mm

galvanized steel profile
vapor barrier

Rockfaçade rockwool soft insulation
membrane fixation

galvanized steel profile
coated metal sheet
joint tape Sarnafil TU 233
edge fixing with welded seam
SARNAFIL FPO TS 77 E membrane mechanically fixed + heat sealed
direction of waterproof grey Sarnafil membrane   gradient > 1%

Rock-up rockwool insulation
fixation with anchor and washer
concrete slab

SECTION - 1/20

Roof Detail 1

172 | Art Museum

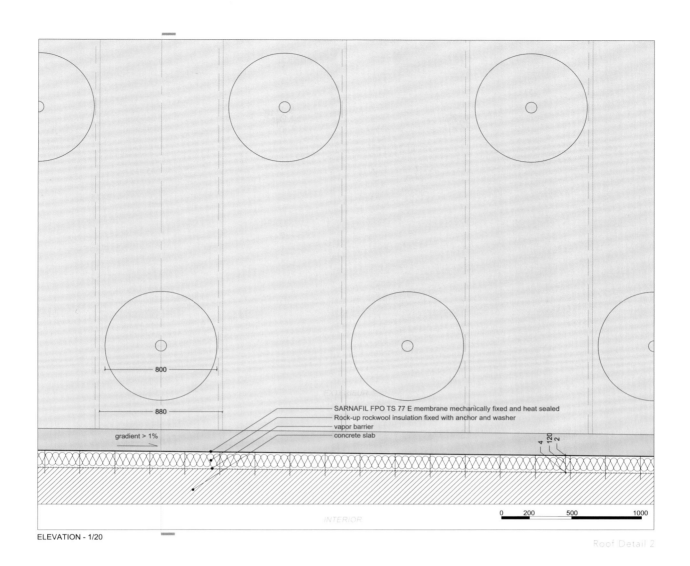

ELEVATION - 1/20

SARNAFIL FPO TS 77 E membrane mechanically fixed and heat sealed
Rock-up rockwool insulation fixed with anchor and washer
vapor barrier
concrete slab

Roof Detail 2

Keywords

Stripped Wooden Cubes
Cost-effective Reconstruction
Meeting Point

Auditorium del Parco

Location: L'Aquila, Italy
Client: Provincia Autonoma di Trento
Architectual Design: Renzo Piano Building Workshop
In Collaboration With: Atelier Traldi, Milan
Design Team: P.Colonna (associate in charge); C.Colson, Y.Kyrkos (models)
Consultants: Favero & Milan (structure and services); Müller BBM (acoustics); Franco Giorgetta (landscape);
GAE Engineering (fire prevention); New Engineering (security); I.T.E.A. (site supervision)

Three striped wooden cubes in the grounds of the medieval castle have provided a temporary concert hall for L'Aquila, bringing life and music back to the city that was hit by a devastating earthquake in 2009.

Among the many buildings destroyed in L'Aquila, capital of central Italy's Abruzzo region, was the auditorium, and this had a significant impact on a city with a fine musical tradition. Restoration and rebuilding work is a long and slow process. For this reason, and to provide a focus and help reanimate the old city center that was largely abandoned after the earthquake, RPBW, with the financial support of the Provincia Autonoma of Trento, designed a flat-packed wooden temporary auditorium. It was constructed in just eight months with the help of local students.

Three stripy, colorful wooden cubes are grouped around a public piazza. Wood is an obvious material for a temporary structure but here at L'Aquila it was also chosen for its acoustic properties. Larch from Trento was famously used by Stradivarius and other 17th-century master lute-makers, and here larch panels have been used to create an auditorium that will resonate like a giant musical instrument. Timber construction is also known to be significantly earthquake resistant, which makes the concert hall a useful prototype for cost-effective reconstruction in L'Aquila's city center.

The stripes of color that animate the three cubes were introduced initially as a code to differentiate between the many cuts of larch slats of different sizes and thicknesses that would be required to build the wooden facades. These colors, which retain this significance, were kept as a part of the final design.

The auditorium itself is inside the largest cube, and is tipped forward at 45 degrees to provide the slope for the raked seating inside. With 240 seats, it matches the capacity of the damaged city concert hall and can accommodate up to 40 musicians.

The walls' raw wood surfaces are hung with a series of acoustic panels orientated towards the audience to reflect sound inside the auditorium.

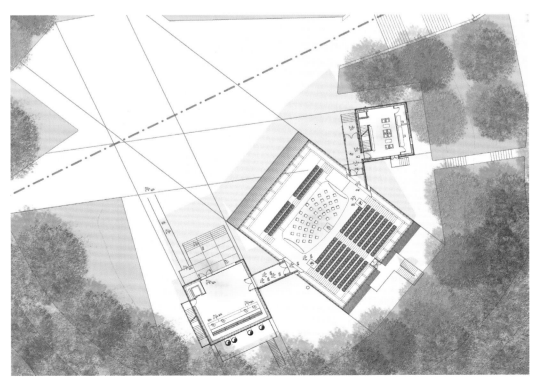

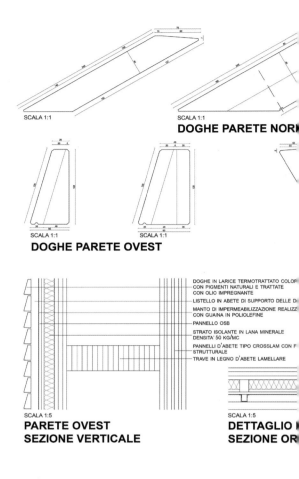

DOGHE PARETE NOR

SCALA 1:1 SCALA 1:1

DOGHE PARETE OVEST

- DOGHE IN LARICE TERMOTRATTATO COLOR CON PIGMENTI NATURALI E TRATTATE CON OLIO IMPREGNANTE
- LISTELLO IN ABETE DI SUPPORTO DELLE D
- MANTO DI IMPERMEABILIZZAZIONE REALIZZ CON GUAINA IN POLIOLEFINE
- PANNELLO OSB
- STRATO ISOLANTE IN LANA MINERALE DENSITÀ 50 KG/MC
- PANNELLI D'ABETE TIPO CROSSLAM CON F STRUTTURALE
- TRAVE IN LEGNO D'ABETE LAMELLARE

SCALA 1:5
**PARETE OVEST
SEZIONE VERTICALE**

SCALA 1:5
**DETTAGLIO
SEZIONE OR**

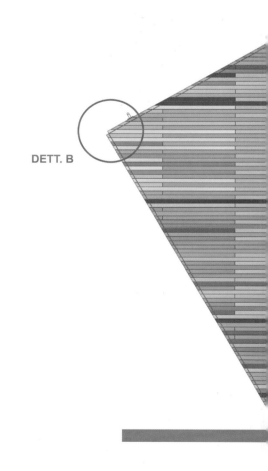

DETT. B

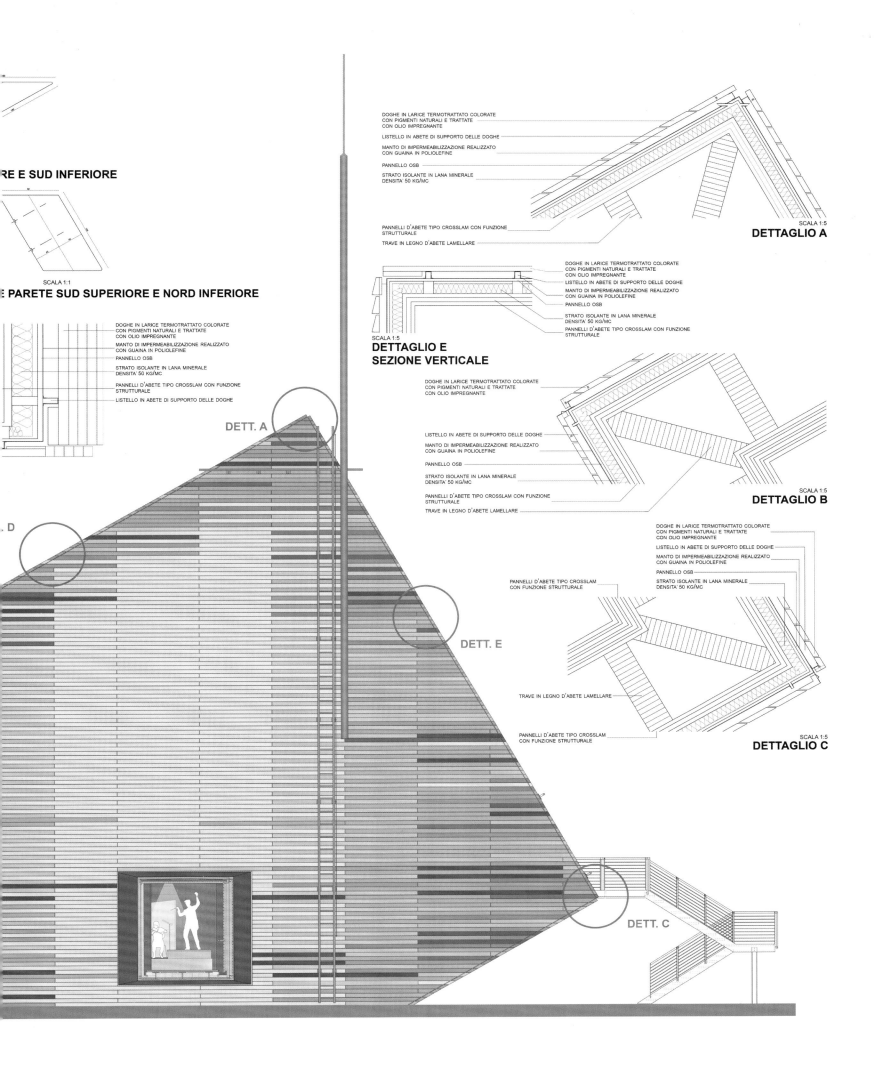

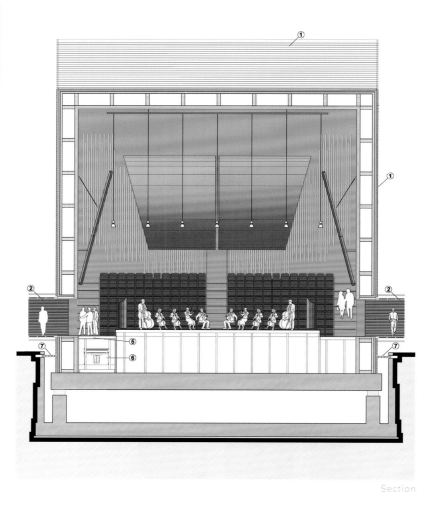

Section

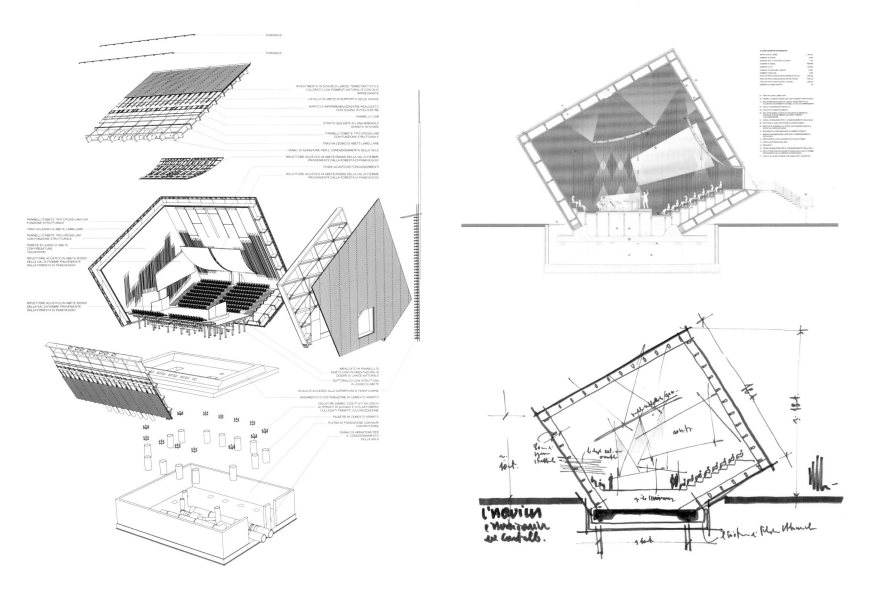

The cube to the right of the auditorium is for the visiting public – a "foyer", with amenities such as box office, cafe and toilets. The cube on the left is for performers: a green room and dressing rooms. A series of glazed corridors link the three structures together. The auditorium is accessed through the foyer via a glazed staircase and elevated walkway.

A public piazza with facilities for outdoor events and performances completes the complex. One of the fundamental aims of the project was the creation of a public space for people to congregate and meet, something that was made very difficult after the earthquake, and recognizing the importance of the spaces between buildings in a city, not just the buildings themselves.

Ninety trees have been planted near the auditorium to symbolically replace the timbers used in the construction of this project. In principal however, this is not a permanent site for the concert hall, since it was always conceived as a temporary structure that can be relocated as required.

Keywords

History
Flexibility
New Beauty

LA FABRIQUE, laboratoire(s) artistique(s) et centre culturel à Nantes

Location: Nantes, France
Client: City of Nantes / Samoa/
Architectual Design: Tetrarc architects
Building Surface: 7,222 m²
Project Manager: Michel Bertreux
Project Director: Rémi Tymen
Project Study: Guillaume Blanchard
Program: Concert hall 1200 seats/ concert hall 400 seats/ 16 recording and training studios/ spaces for digital experimentation/ offices/ welcoming public spaces (cafe, restaurant, bar).
Photography: Stéphane Chalmeau

Program

Design an equipment dedicated to contemporary and emerging music, assembling together two performing show halls (1200, 400 places), 16 re-cording and training studios, spaces for digital experimentation, offices, and welcoming public spaces.

Project

On the "île de Nantes", La Fabrique takes place on one side of the previous Dubigeon warehouse, near the Alstom warehouse (the upcoming center of the "Quartier de la Création") and the old shipyard turned to great gardens.This context gave Tétrarc the opportunity to gather the recalls of a place abandoned by History and remind us that Culture is continuum made of sedimentation and fertile rediscovery.

History of places made of industrial warehouses. History of the techniques that built them, when science and aesthetic meet up to create a new beauty. History of the ships built with the strength skills and talents of the workers. History of artists whose eyes felt on this universe, from Monet, painter of the brand new steel structures, to the Becher who used to forget metallic shapes.

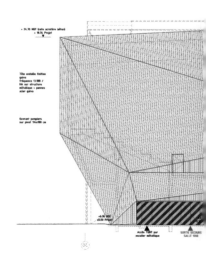

Again, History of the industrial production organization which enable to place under the same roof of an experimental workshop, an assembly line, a rectification workshop, and an export site ; all those as a living picture of the concept of flexibility.

The project is made of 3 distinct elements. The 400 Hall (with offices in it), the 1200 Hall: two parts connected by a public Hall inserted in the concrete post frame of the Halle Dubigeon. And the studios take place in a volume suspended above an old air-raid shelter.

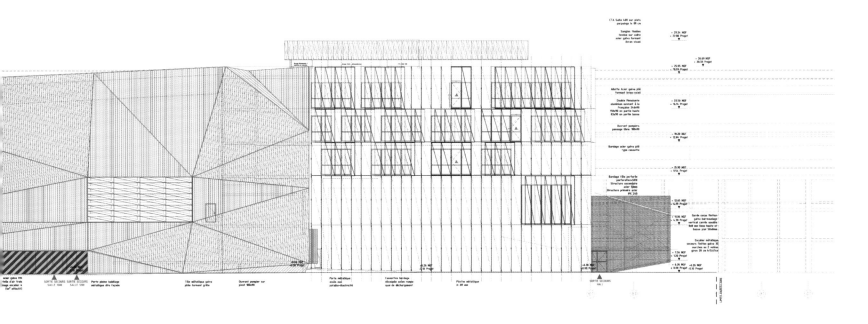

185

Keywords

Superb Facilities
Robust Flexibility
Multifunctional

Birmingham Ormiston Academy

Location: Birmingham, UK
Client: Lend Lease
Architectural Design: Nicholas Hare Architects LLP
Landscape Design: Nicholas Hare Architects
Interior Design: Nicholas Hare Architects
Gross Internal Floor Area: 8,990 m²
Photography: © Alan Williams photography

Birmingham Ormiston Academy specializes in Creative, Digital and Performing Arts and is located in the heart of the Eastside Learning Quarter at the hub of the digital and creative industries in the West Midlands. Designed by Nicholas Hare Architects LLP, the Academy offers superb facilities and robust flexible infrastructure for academic, vocational and extra-curricular activities, including a theatre, TV studio, radio suite, editing facilities and a recording studio sponsored by the BRIT Trust. In addition to four dance and six rehearsal spaces, there are also general academic teaching spaces.

The Academy provides a center of excellence for training students for the creative industry. A vibrant atrium foyer space at the heart of the building serves the key performance spaces. The theatre will be used throughout the year for public performances. An external performance space is complemented by a 13 x 5m screen on the theatre facade that advertises and celebrates the digital media specialism and inner workings of the Academy to the outside world.

Image depicting the council's vision for Eastside © Birmingham City Council

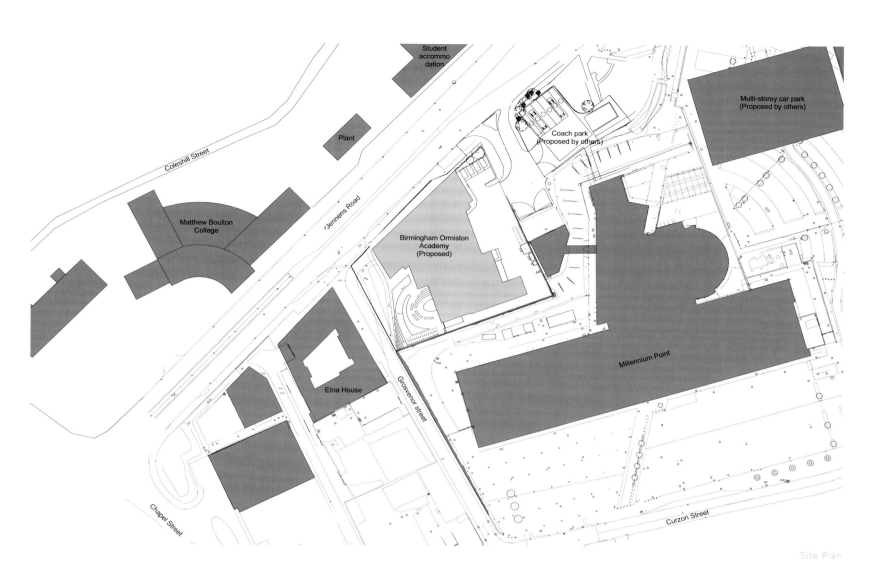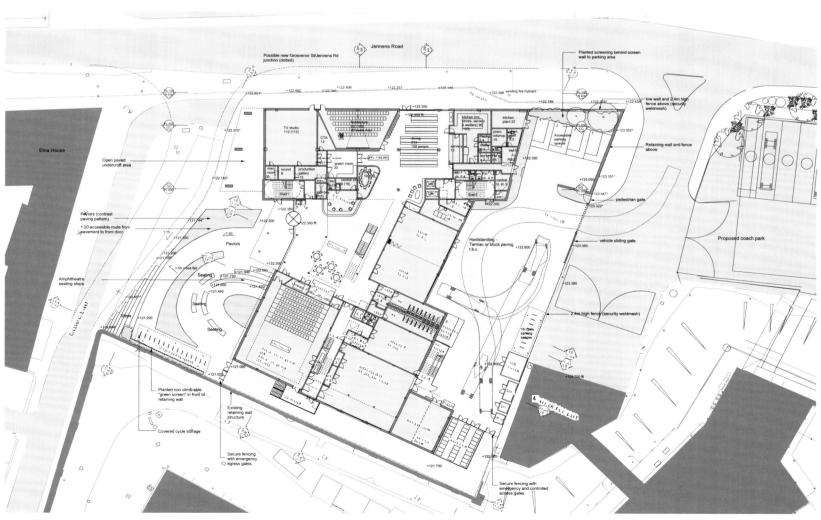

Section

194 | Art Museum

Section A-A

Section B-B

Section C-C

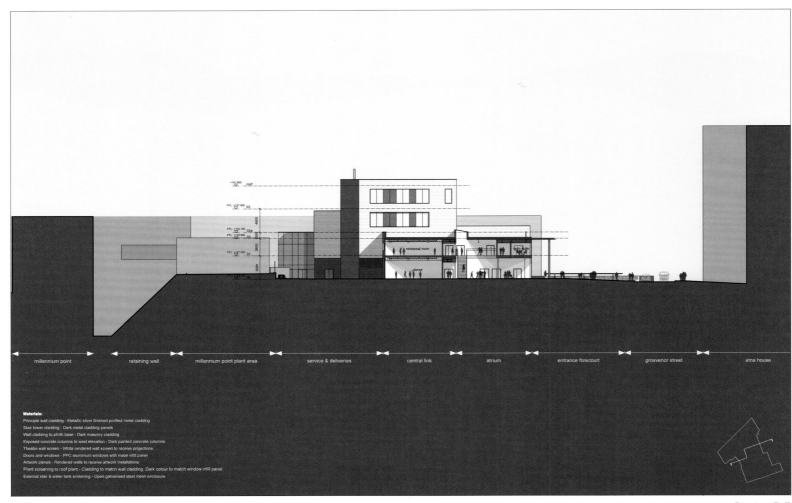

Section D-D

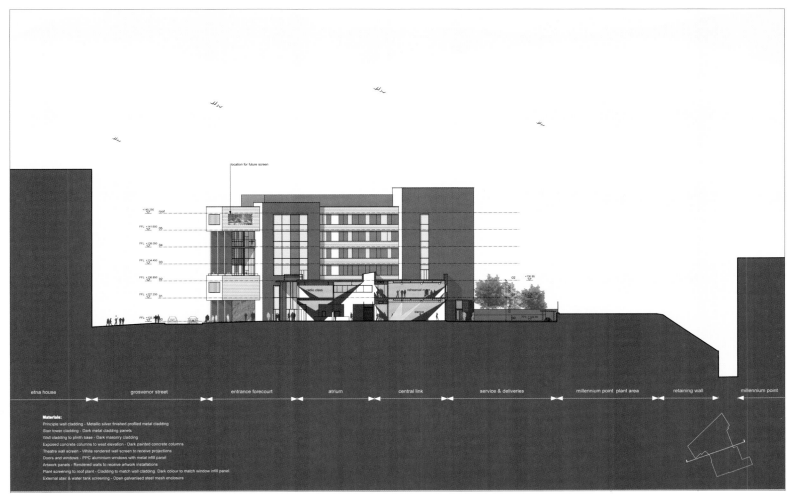

Section E-E

197

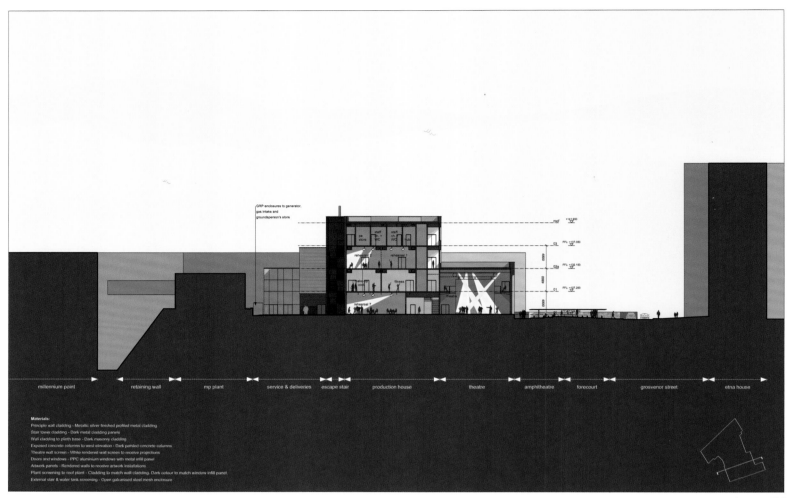

Section F-F

Art Building

Art Performance
Cultural Value
Artistic Appreciation

Keywords

Interactions
Pleated Roof
Baroque Posture

Théâtre 95

Location: Cergy-Pontoíse, France
Architectual Design: GPAA (gaëlle péneau architectes associés)
Site Area: 10, 000 m²
Floor Area: 3000 m² auditorium
Photography: 11h45

The Challenges of the Project
The contractor's definition of the project is articulated around a close link between reshaping the Théâtre 95, outlining a urban morphology, and finding free spaces in the urban island. The theatre's extension is framed by a dialectic vision of interactions between town and theatre: the theatre stretches beyond its limits while the town must find in it a porous space, a space of journeys, echoes, dreams, encounters and everyday life, devoted to contemporary modes of expression and deliberately placed at the very heart of the town.

A Peculiar Face-off
The Théâtre 95, opposite the prefecture, is housed in the third building to be built in the new town which arose in the 1970s. This emblem of the town's history was once the home of Cergy Pontoise school of architecture and urban planning before becoming an arts school and then being transformed into a theatre. The extension is a complex project which is framed by more general considerations about emerging social and urban developments and cultural practices. The aim is to invite the wider public to discover new strategies to reinvent the town.

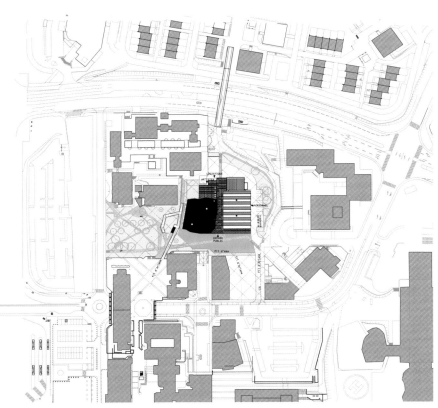

203

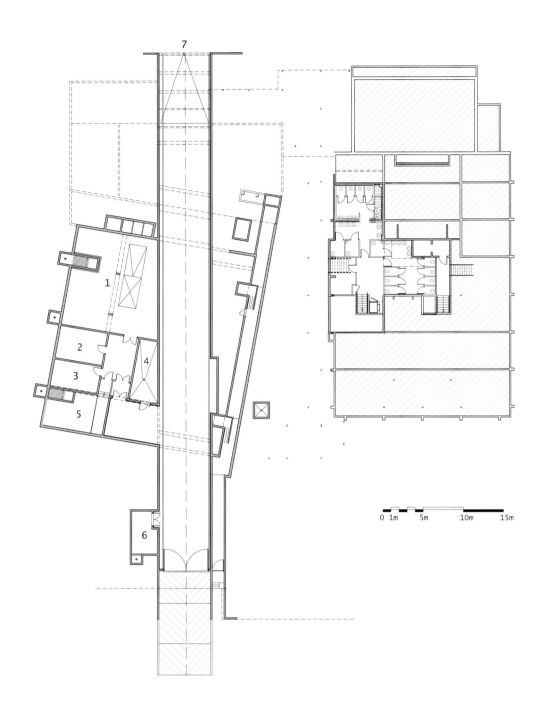

1 local ventilation
2 sous-station chauffage
3 local eau
4 dépôt technique
5 cuve de récupération EP
6 transfo
7 accès vers bâtiment existant

0 1m 5m 10m 15m

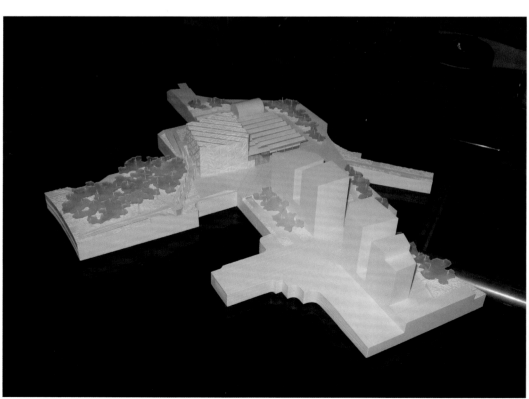
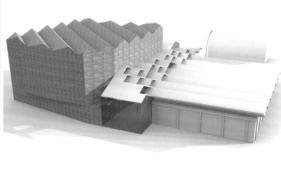
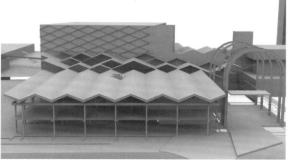

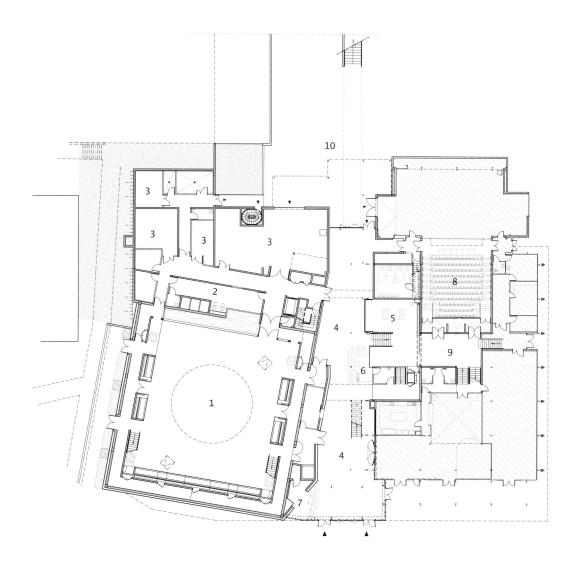

1 salle
2 loge de proximité
3 dépôts
4 atrium
5 librairie
6 bar
7 accueil
8 salle exsitante
9 hall
10 aire de service

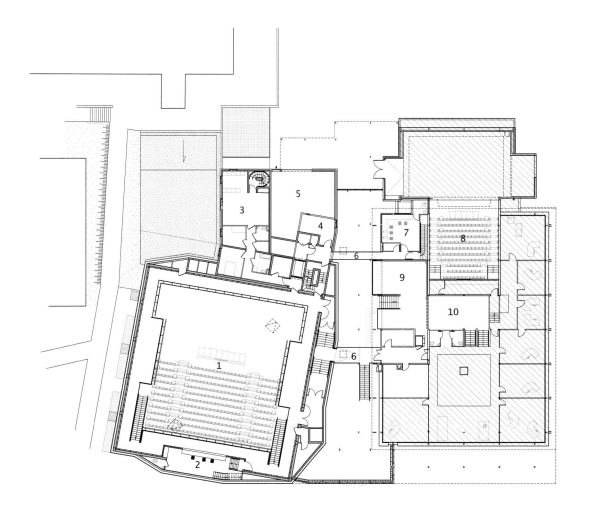

1 salle
2 régie
3 logement gardien
4 bureau techniciens
5 vide sur dépôts
6 passerelles dans l'atrium
7 loge collective
8 salle exsitante
9 librairie
10 hall

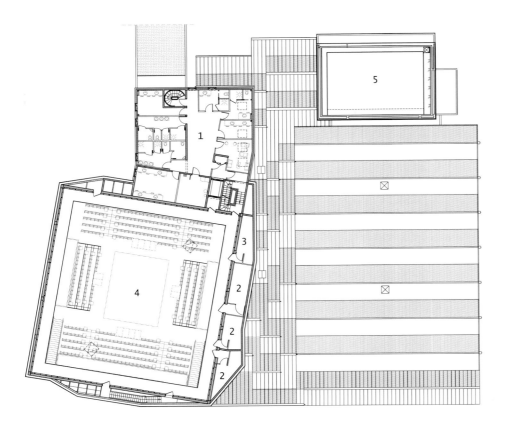

1 annexes artistiques
2 dépôts
3 local onduleur
4 salle
5 vide sur scène

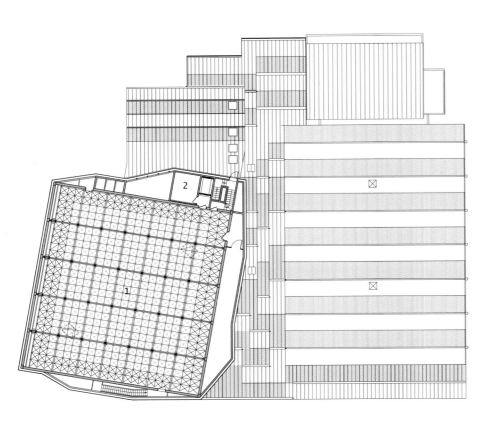

1 gril
2 local ventilation

The building's pleated roof is the first component which strikes the visitors' eyes: this is the outline which the extension has borrowed to link old and new. The connection consists in a "semi public" hall-atrium, which follows the "Fil D'Ariane" - a public footpath which winds its way without interruption from the South-East to the North-West of the town, and is thus "integrated" into the building.

The existing pleated outline of the roof is continued in the hall-atrium volume, where it transforms into juxtaposed strips which create shafts of light entering the hall. The pleats are also echoed in the new auditorium, facing South, creating a new rhythm which emphasizes the choice of erecting the new structure out of line with Cergy's traditional orthogonal grid. The pleated outline has become the "crown" which is found in the volume of the new auditorium.

The new volume rises in an almost baroque posture, as if in confrontation with what is already there: the existing building conserves its identity, the atrium linking it to the new, setting up a "face-off" relationship between two visions which mix, stand in opposition and join together in a boldly chaotic statement. The new extension will house a "flexible" 400 seats auditorium: the volume, which includes stage and technical areas, is blind, and covered with golden scales which bring light to a fairly colorless urban environment.

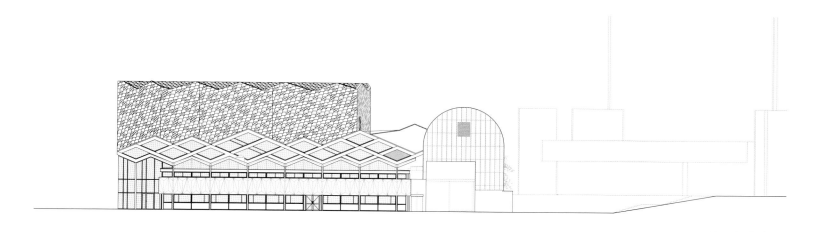

Facade Sud-ouest

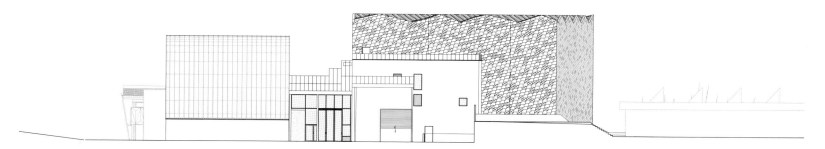

Facade Sud-est

Facade Nord-ouest

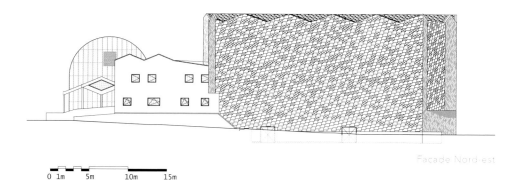

Facade Nord-est

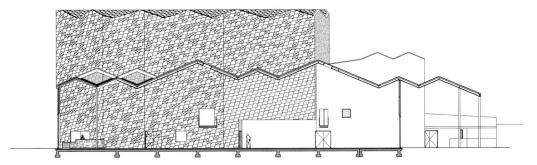

Coupe Sur L'atrium Vers La Nouvelle Salle

Coupe Sur L'atrium Vers L'existant

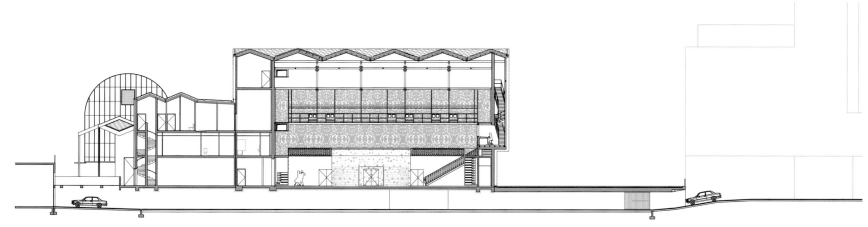

Coupe Longitudinale Sur La Salle Et La Regie

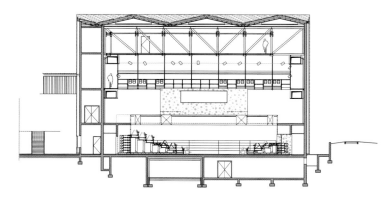

Coupe Transversale Sur La Salle Vers La Regie

The choice of materials was based on a certain number of considerations which include not only questions of maintenance, longevity, environment, energy efficiency, but also appearance and style. Copper offers a response to all these questions. A performance space is also a strong presence at the heart of a town: it is a playful space, designed for culture and leisure which, like a magic lantern, must shine and draw all eyes towards it. Thus the envelope is made up of smooth, diamond-shaped scales, made from a copper aluminum alloy lending them a golden shade which will fade very little over time, and also contributing to lighting the hall-atrium. Within the auditorium, the space becomes more technical: Shaped as a cube to which the lighting and sound box, storage areas and dressing rooms are attached, it is equipped with retractable seats on 4 sides, making any seating layout possible, and a technical grid tailored to the dimensions of the auditorium. The audience enters via a footbridge on the first level. These technical characteristics are supplemented by details which create a welcoming and baroque atmosphere: the wooden acoustic panels are carved with motifs inspired by the orchard behind the theatre, the wrought iron balustrades hark back to theaters of old, like the iron-work of Parisian balconies, the deep purple hues declined over the walls and seating which are lightened with touches of red here and there.

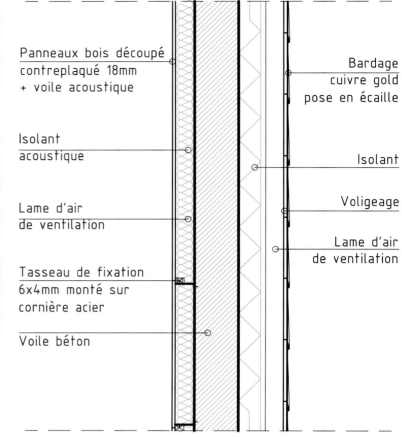

Panneaux bois découpé contreplaqué 18mm + voile acoustique

Isolant acoustique

Lame d'air de ventilation

Tasseau de fixation 6x4mm monté sur cornière acier

Voile béton

Bardage cuivre gold pose en écaille

Isolant

Voligeage

Lame d'air de ventilation

Detail Paroi

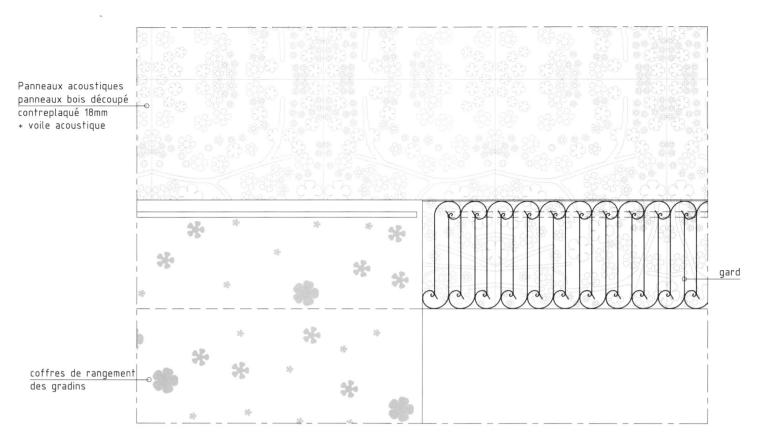

Panneaux acoustiques panneaux bois découpé contreplaqué 18mm + voile acoustique

gard

coffres de rangement des gradins

Detail Decorum

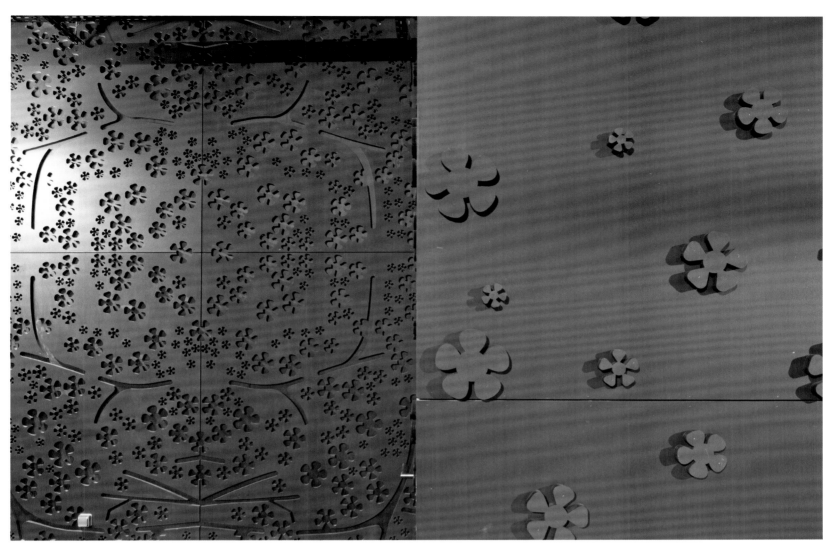

Keywords

Cultural Quarter
Optimal Contact
Theatrical Feel

Theater De Nieuwe Kolk, Assen

Location: Assen, the Netherlands
Client: BAM Utiliteitsbouw Groningen
Architectual Design: Greiner van Goor Huijten Architects

The project not only involved replacing the theatre, but also constructing a multifunctional building on the De Kolk site. This building was to feature a number of uses: two theatre auditoriums with 850 and 250 seats and all associated facilities, five cinemas, a library, hospitality facilities, a Centre for Visual Arts, foyers, apartments, bicycle storage, a parking garage for 500 cars, a space for Biblionet Drenthe and commercial offices. The project was one part of a competition, and the design was created by Greiner van Goor Huijten Architects together with De Zwarte Hond Architects. The tasks were then divided, and Greiner van Goor Huijten Architects became the lead architects for the theater section.

The design for the theater was based on creating optimal contact between actors and audiences. The theater's shape, with two balconies, means the audiences "embraces" the actors. The small auditorium is a multifunctional "shoebox" with bleacher seating. Side balconies enhance the theatrical feel of the space.

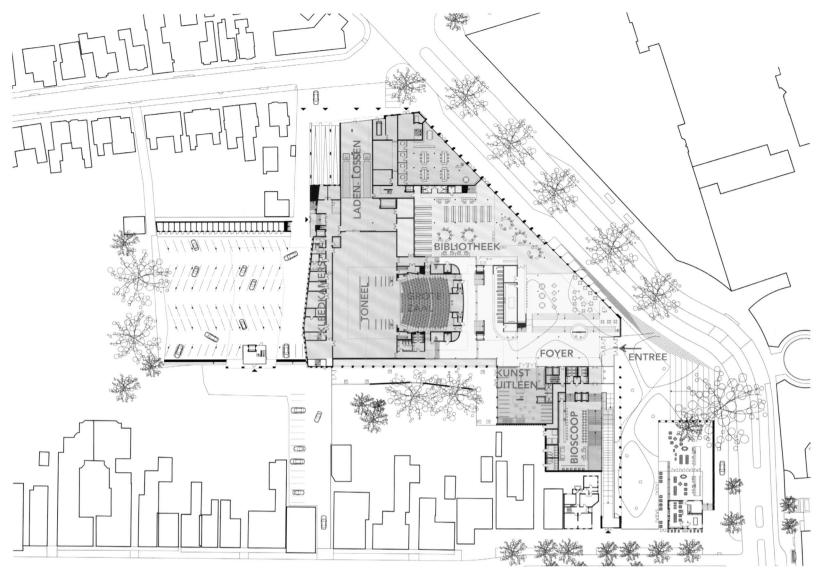

Ground Floor Plan

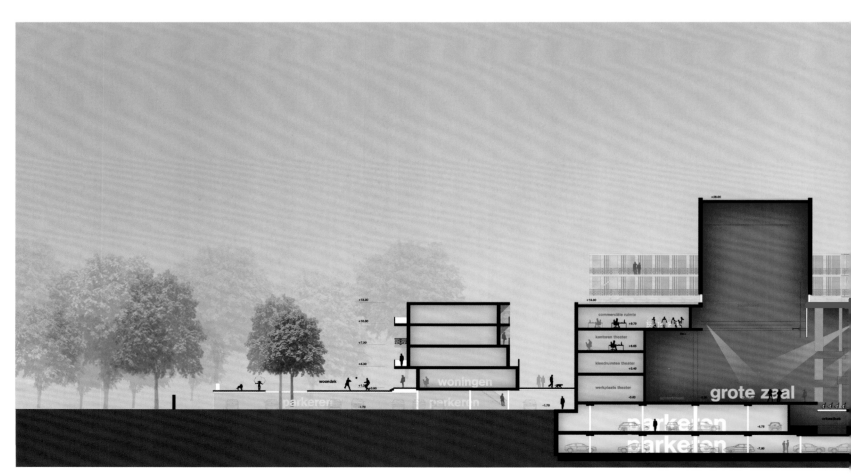

218 | Art Museum

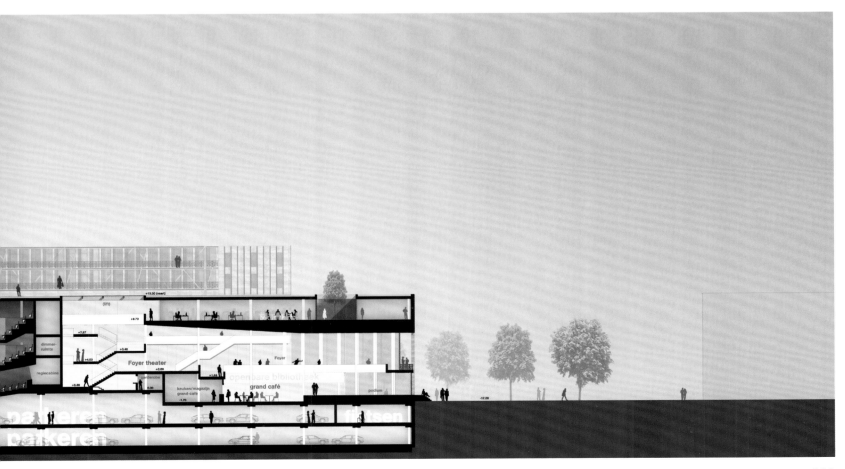

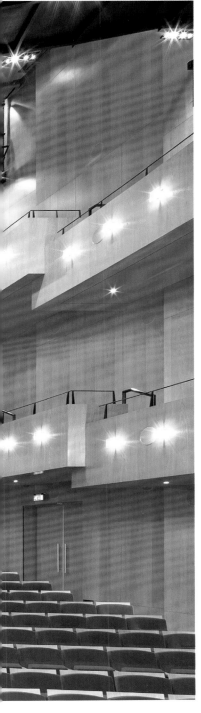

Keywords

Multi-purpose Parallelepiped Urban Gallery

New Theatre in Montalto di Castro, Italy

Location: Montalto di Castro, Province of Viterbo, Italy
Client: Municipality of Montalto di Castro, Italy
Architectual Design: MDU Architetti (Alessandro Corradini, Valerio Barberis, Marcello Marchesini, Cristiano Cosi)
Collaborators: Nicola Becagli, Michele Fiesoli
Structural Engineers: ACS ingegneri
Mechanical And Electrical Engineer: Federico Boragine
Site Area: 10,888 m^2
Roofed Area: 963 m^2
Gross Area: 1,220 m^2
Materials: concrete, wood, alveolar polycarbonate
Photography: Lorenzo Boddi, Valentina Muscedra, Pietro Savorelli

The New Theatre in Montalto di Castro, designed by MDU Architetti, is located in the Maremma Laziale, in the province of Viterbo, in Central Italy. The building is located on the edge of the city, in an area bordering the most consolidated urban fabric and the most rarefied on the urban margins. This position suggests one of the theatre's main vocations: mediation between the two areas of the city. The theatre is a public work, the result of an international design competition launched by the Municipality of Montalto di Castro in 2002 (results published in 2004), and was built in six years. The town administration's goal was to redevelop a former industrial area so it could be used for the construction of a multi-purpose theatre – to host performances, conferences, and recreational activities – which would represent a "driving force for the cultural growth of the community".

The theatre, a parallelepiped with a vertical fly tower, houses – on an area of 963 square metres – the foyer, the auditorium that seats 400, the outdoor arena that seats 500, administration and service rooms, and a car park with around 60 parking spaces. Two elements that helped to shape the project are the nearby Etruscan ruins and the Alessandro Volta power plant. The former, and specifically the base of the Great Temple of Vulci, inspired the parallelepiped monolith, home to the foyer and auditorium; the latter suggests the idea of the glass fly tower whose vertical form becomes a point of reference and a signal point for the territory. "Archaic Etruscan versus the aesthetics of the machine", explains MDU, alluding to a temporal "short circuit" that inspired the architecture and projected it into an imaginary dimension in which history and modernity coexist and confront each other.

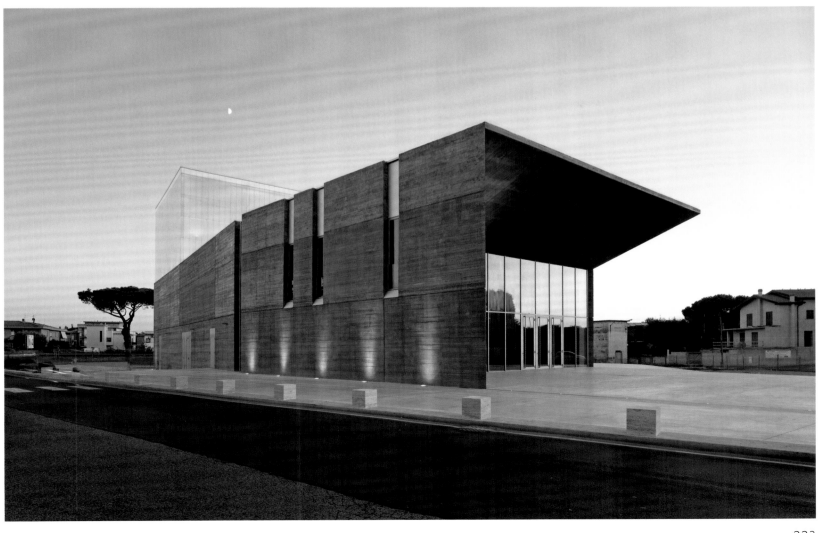

planimetria generale - scala 1:200

Inside, the monolith reveals a deep fissure that runs through its mass from one end to the other: a sign of erosion that creates a fluid and open architecture, a channel of interaction with the city. The theatre is thus ideally traversed by the urban context, becoming a new urban fragment. Along this section the foyer seamlessly flows into the auditorium without filters, creating a liquid space that can be crossed from the entrance as far as the summer arena. The theatre becomes an "urban gallery" that conveys culture and transmits it to the city.

The main materials are concrete for the monolith, strips of wood covering the vertical structures creating a sequence of warm and vibrant curtains, and alveolar polycarbonate for the fly tower, which by day dematerializes becoming indistinguishable from the sky, while at night it lights up like a beacon on an urban scale. The building, one moment through comparison with pre-existing structures and the next through the sign of erosion within it, interprets two topics dear to MDU: the "poetic measuring" of the territory, as defined by

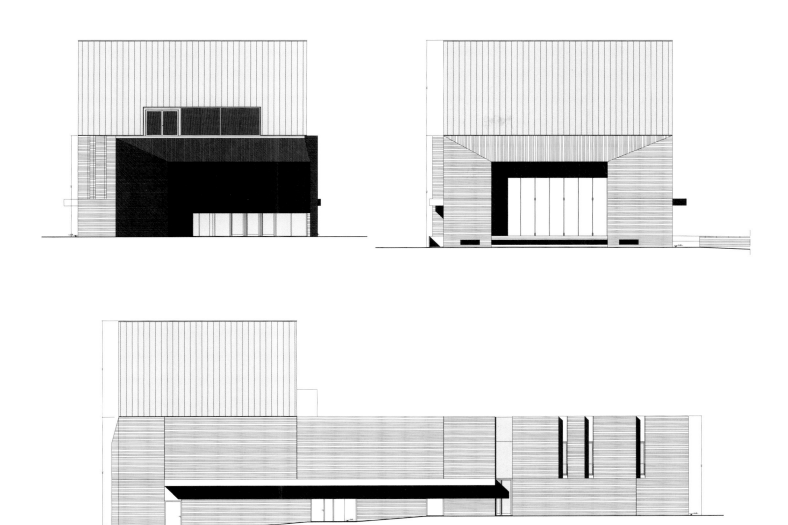

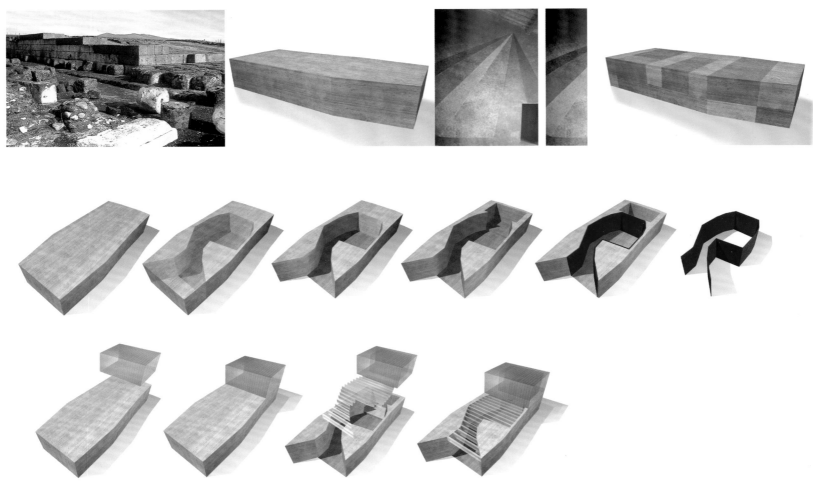

226 | Art Museum

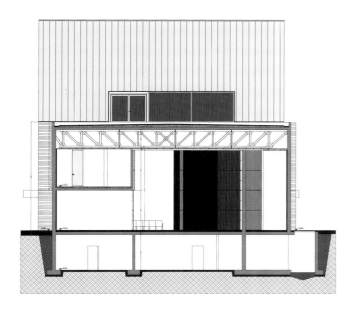
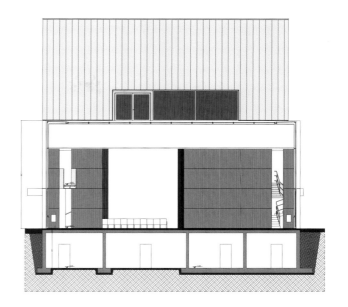
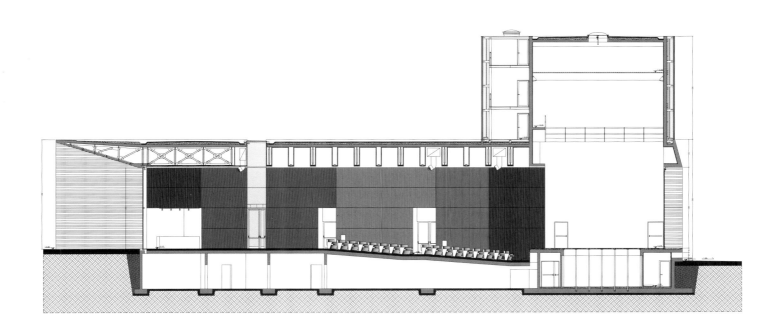

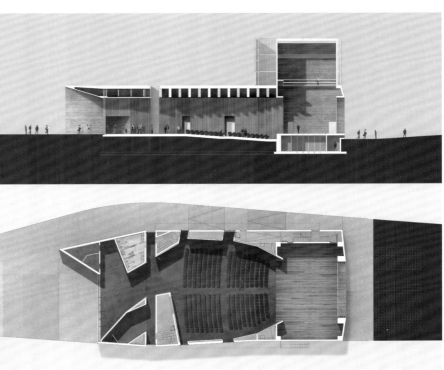

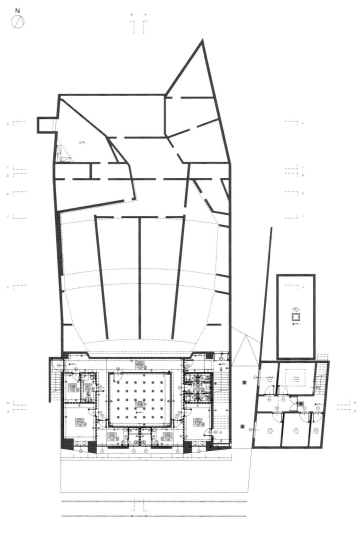
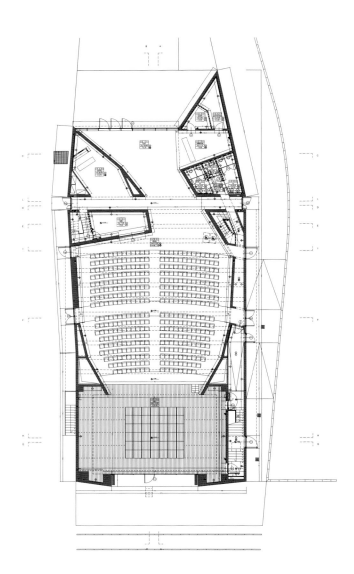
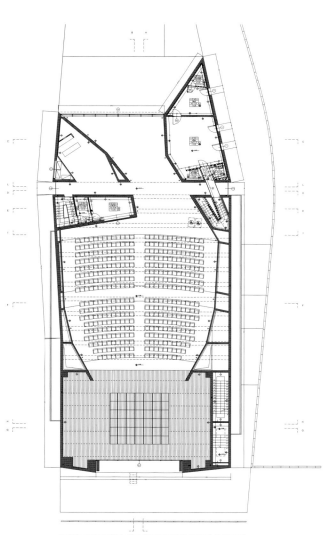
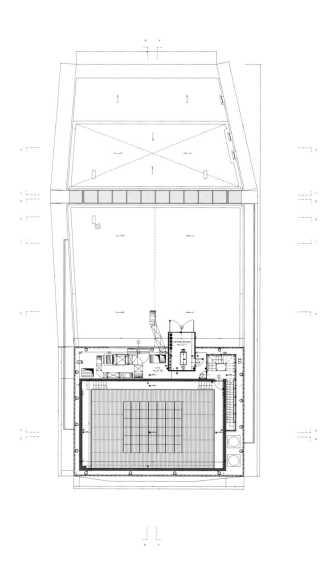

228 | Art Museum

the designers, and the desire to create a place conceived as a "journey of approaching the performance". "The new theatre is proposed as a conceptual model for measuring the territory and at the same time it attempts to express, through architecture, the magic of a theatrical event felt by the audience", adds MDU.

The theatre becomes a catalyst of attention from the entrance, through which the foyer can be seen flowing directly into the auditorium describing a single room, a continuous ribbon made dynamic by the fragmented series of vertical wood panels that guide the spectator towards an atmosphere of awe, wonder and spectacle.

The spatial hierarchies become loose, relax their classifications and succumb to the democratic organization of the space in which the various areas merge into one another. A fluid journey that does

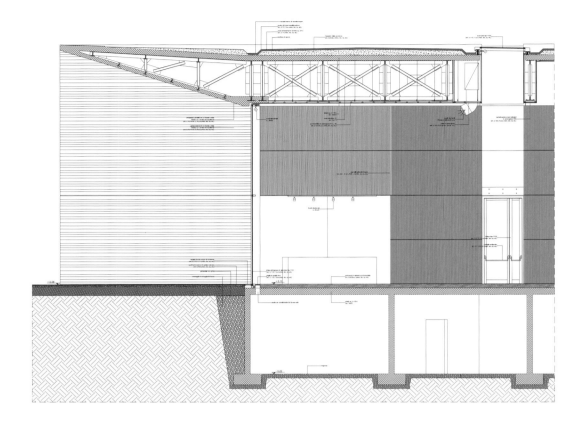

not stop at the stage, rather crossing it to describe an outdoor arena from which it is possible to enjoy theatrical performances even in the summer season, thanks to the transparency of the fly tower.

With its appearance, the theatre offers the city a new entrance, positioning itself as a permeable gate, an ideally traversable place, like a flow that guides its visitors towards the urban landscape. The entrance, with its impressive projecting frame, is the first element that emphasizes the invitation to the city; the welcoming nature of its design embraces the open space in front of it, making it the heart and soul of the piazza. "Everything has been designed to encourage individuals to experience the 'place of the theatre' and the area 'around the theatre' as if they are related", says MDU.

233

Keywords

Multifunctional
Attractive Experience
Diversified Performances

Opera House and
Pop Music Stage
Enschede

Location: Enschede, The Netherlands
Commissioner: City of Enschede
Architectural Design: Ector Hoogstad Architecten, Rotterdam, The Netherlands
Designer: Jan Hoogstad
Users: Music school Twente, Popstage Atak, Podium Twente, The Orkest van het Oosten,
	the Nationale Reisopera and The ARTez conservatorium.
Surface Areas: 18,000m^2
Photography: Jeroen Musch

This multifunctional music and theatre center—comprising three halls, office spaces, studios and classrooms—accommodates a variety of users from all areas of Twente's music scene: Podium Twente, the Orkest van het Oosten, the Nationale Reisopera, Muziekschool Twente, the ArtEZ Conservatorium and Poppodium Atak. A number of facilities are used collectively. The music school in the complex has a close but separate relationship with the theatre; students share the building with professionals. The communal space has been arranged to generate a high level of synergy.

Buildings featuring theatrical stages are among the most complex design projects. The main reason is that such venues are expected to be extremely flexible in order to accommodate a wide diversity of performances. The largest auditorium is technically equipped to accommodate plays and all sorts of dance and musical performances. The technical equipment and interior of Hall B ensure that, on a single night, this big pop venue can be magically transformed from a concert hall for all kinds of amplified music into a nightlife temple for DJ-hosted dance parties. In addition, Hall B can be used for small theatre performances and shows that require no stage. Backstage, the three halls share a spacious and efficient loading area, and the extra space thus created facilitates the assembly and dismantlement of stage sets, even in the case of complex productions. The individual venues are free to programme independently: thanks to box-in-box constructions and architectural expansions, none of the three is inconvenienced by the others.

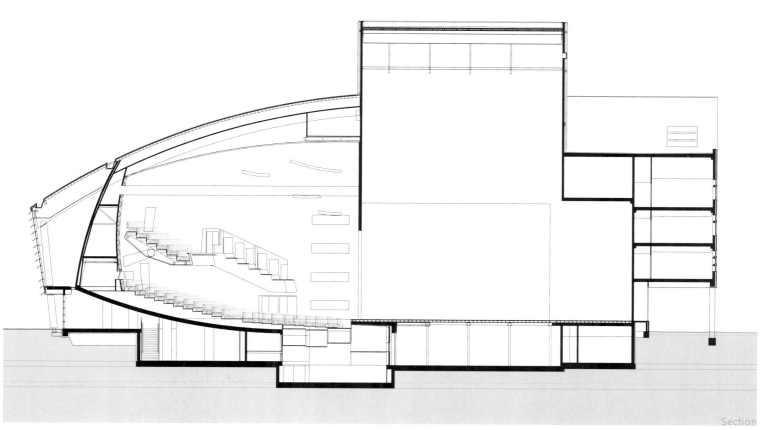

Section

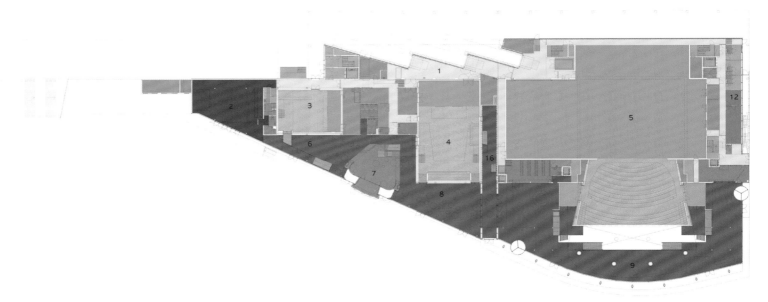

First Floor Plan

A building for performances can be successful only when it is functionally perfect and makes performers feel good. A good theatre enhances the player's performance. The concept shows three zones: backstage zone, play zone, audience zone. Each zone has its own atmosphere. The backstage area is bright, stylish and sometimes homely. It offers a comfortable working environment that performers find pleasant and are glad to come back to, combined with a well-oiled machine that offers a wide range of options.

The audience finds the experience perhaps even more important, and each target group has its own wishes. Opera-goers have different expectations than teenagers who come here to dance. The building plays an important role in such experiences, and all kinds of resources - color, material, texture, decoration - are employed to create an attractive experience. At night, a sophisticated lighting system is able to change the architecture of the three foyers, which form an entity during the day, into individual interiors with distinctive ambi-

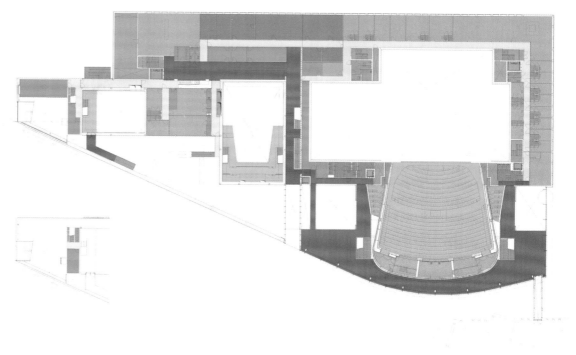

Second Floor Plan

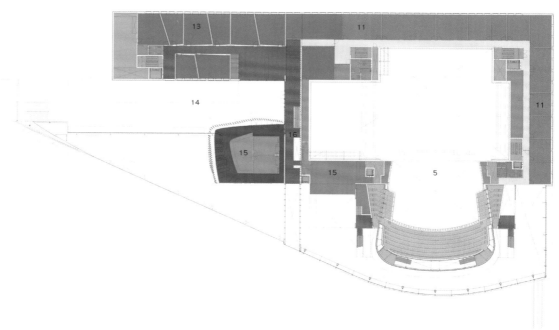

Third Floor Plan

ances. Whereas the theatre foyer often provides a romantic and festive environment with an open character, and its accompanying auditorium set among various levels like a jewel, the foyers of pop venues are usually more abstract and introvert, with lighting that visually lowers the space and directs attention to floor and skirting board. The foyers of the so-called Muziekkwartier can also be interconnected as a sequence of three interiors linked by ornament on the rear wall. In addition, a footbridge connects this area to the adjacent music center. The design allows Enschede to organize a huge music festival: a bonus afforded by joint accommodation.

From first to last, the theatre owes its existence to the performance: the encounter between players and their audiences. Besides top-notch technology on and around the stage and pleasant foyers, the atmosphere within the auditoriums is decisive. The largest of the three has a warm and somewhat classic character achieved with red fabrics and dark woods. The basic form is shaped like a shell. This makes the double-curved rows of seats seem to hug the stage so to say. The small Hall C has the snug, warm feel of a club, with parquet floors, wood panelling and prefab concrete panels with mouldings. The larger Hall B is dark, rather neutral and yet rugged, thanks to a combination of perforated Cor-Ten steel sheet, the same type of prefab concrete panels, and a black synthetic floor.

This rich selection of required facilities was cleverly assembled within a striking complex on a cramped urban site nearby the railway. With a huge outdoor staircase that faces the station and a slanting green-copper roof above a glass facade on the side bordering the city center, the Muziekkwartier beckons passers-by approaching from all directions. It is, moreover, contributing to a logical and attractive routing through the city.

Keywords

Staggered Glazed Elevations
Warm Natural Materials
Aluminum Curtain Walls

Pier K Theatre and Arts Centre

Location: Nieuw-Vennep, The Netherlands
Architectural Design: Ector Hoogstad Architecten, Rotterdam, The Netherlands
Designer: Joost Ector
Surface Areas: 2,750 m^2
Photography: Jeroen Musch, Kees Hummel

The arts building houses Pier K, the cultural organization responsible for promoting activities and art education right across Haarlemmermeer. The design challenge was to find a balance between the desire for a prominent building on the one hand, and on the other the need for a warm, welcoming and low-key edifice. In other words not a cultural temple but rather a club house for culture. The complex programme extends over three floors. All the rooms, the grand café and all the service functions are situated on the ground floor. The teaching rooms for music, dance and the visual arts are on the first and second floor. Staggered glazed elevations point-up the site of the entrance area and of the terrace. Daylight penetrates through the glazed entrance elevation, but also through the central well of clear space that runs all the way through the building.

In the interior, warm natural materials have been combined with vividly colored walls. The facades are made up of fully glazed aluminum curtain walls, with western red cedar timber panelling, and are largely clad in slate. The slate facades have been furnished, seemingly at random, with windows of varying dimensions. From the inside these windows provide a range of interesting views of the surroundings and, conversely, function as shop-window for the activities in the building. During the evening hours, when the building will be intensively used, the building turns into a glowing coal in the new heart of Nieuw Vennep.

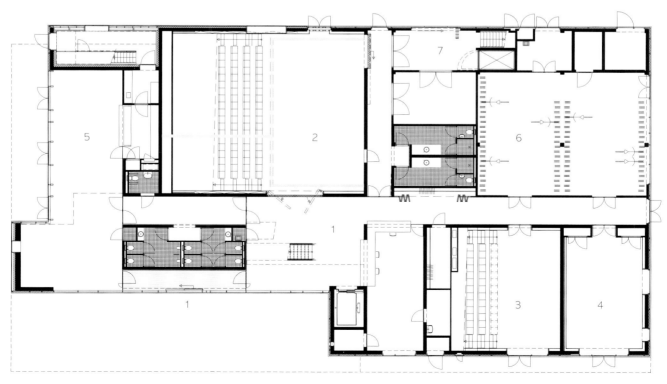

First Floor Plan

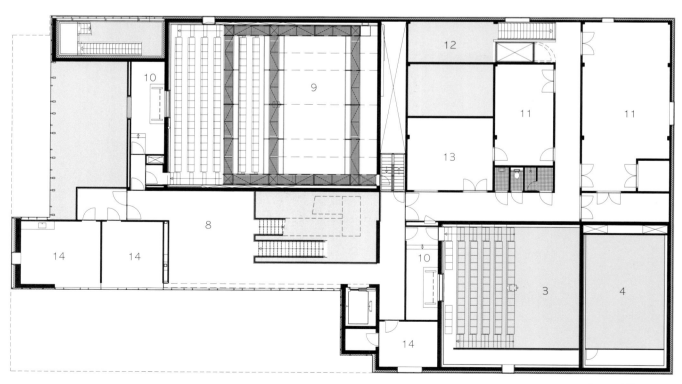

Second Floor Plan

242 | Art Museum

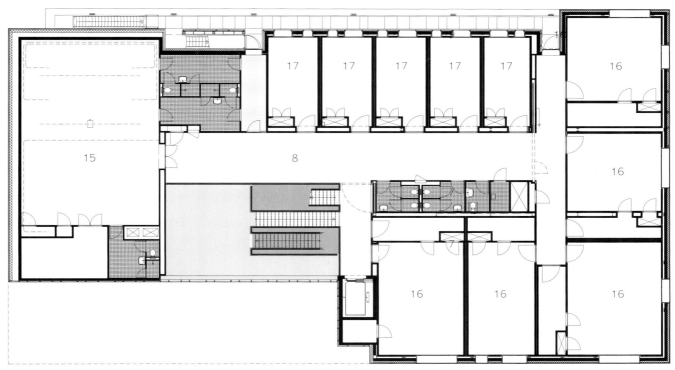

Third Floor Plan

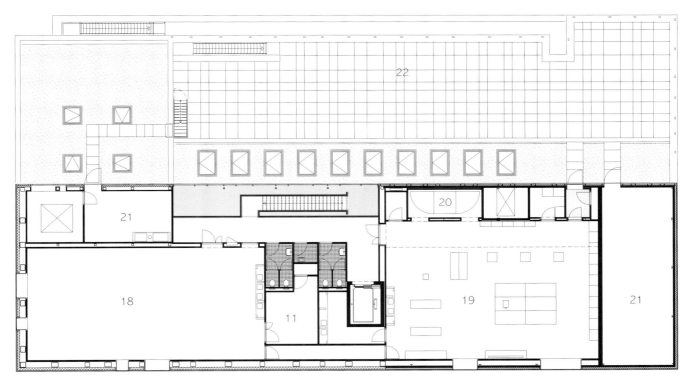

Fourth Floor Plan

1 entrance
2 popstage
3 auditorium
4 mixed use
5 grand cafe
6 bycycles
7 traffic
8 foyer
9 technique
10 stage-management
11 storage

12 plateau
13 studio
14 office
15 dance studio
16 music studio
17 music studio
18 art (painting) studio
19 art (sculpture) studio
20 ceramics
21 installations
22 roof terrace

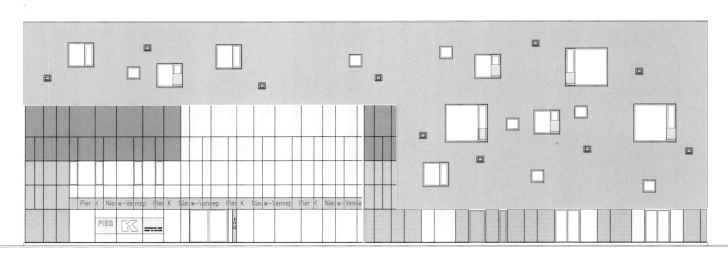

Section 1

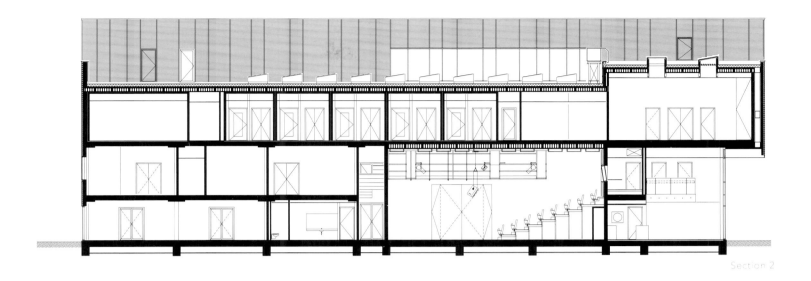

Section 2

Keywords

Redevelopment
Redesign
Reinterpretation

Hamer Hall

Location: Southbank VIC 3006 Australia
Architectual Design: ARM Architecture
Photography: John Gollings, Peter Bennetts

The underlying ideas behind the design came from the strength of the conflicting thematic by Roy Grounds and John Truscott. The commission comprised a review of the master plan for the whole of the Arts Centre precinct its relationship to greater Southbank. The mission, gradually refined through conflicts and agreements with the various client participants was to open up Hamer Hall, integrating it with the river and make it a 21st Century venue. Underlying the job was to repair its standing as a major civic building. The building represents two artistic discourses in Melbourne: the architectural story of the castle and the mine: the theatrical story of the palace and the cave of jewels. The architects took these two narratives, reinterpreted them and added the architects' voice to the story of Hamer Hall. The new riverside section is inspired by a new geometry.

The redevelopment was approved by Heritage Victoria. Throughout the redevelopment the architects took great care to preserve many unique heritage aspects of the Hall, retaining Roy Grounds' base architecture and as much as possible of John Truscott's original interiors.

The redevelopment repairs HH's performance acoustics, to world standards, installs state of the art stage technology and back of house facilities. The patron amenity is completely redesigned, with increased foyer area, toilet numbers and lounges with exterior modifications that create a more outward facing venue to make it more accessible and inviting to the public.

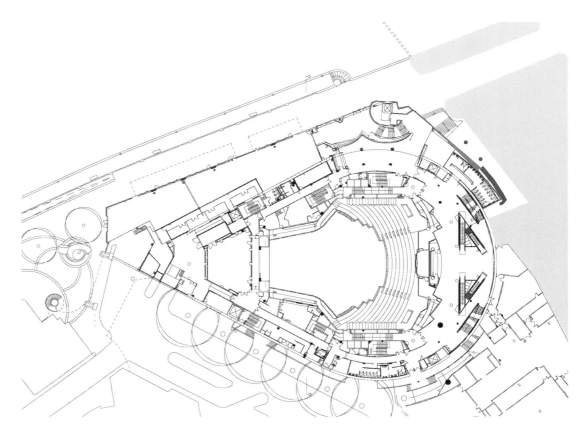

Site Plan 1

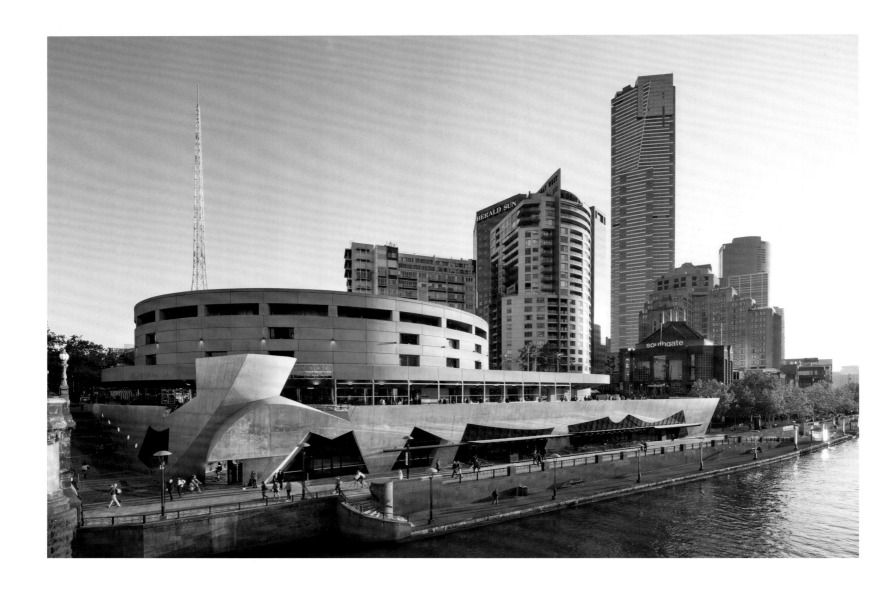

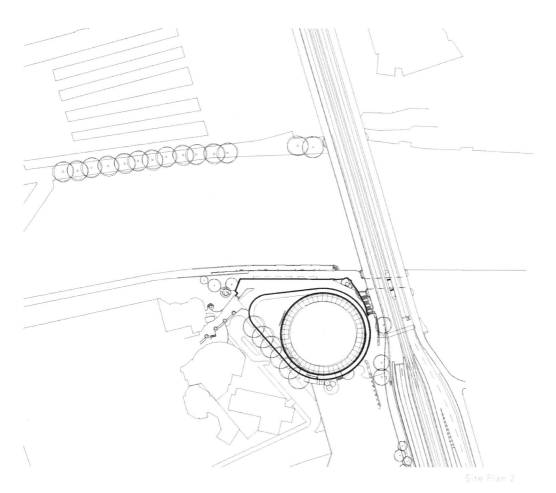

The existing built form of the drum has been retained but radical changes were made to the riverside section of the precinct. The new form draws its influences from Grounds' muse, the Castel San Angelo, Rome. Built in off-form concrete it recalls the rusticated base of the Roman ruin. The layout reinforces the existing ant tracks around the building.

The commission was to rethink the master plan completed earlier by FJMT. The funding for the project had been primarily focused on high level ambitions for re-branding Hamer Hall and repairing theater technology and poor access. Quickly the team had honed in on the main programmatic challenges to be addressed. These included improved integration with the Hall's surroundings, especially St Kilda Rd and River side, substantial Hall acoustic repair, substantial operating and technical improvements, a third revenue stream through new F&B spaces and improved patron amenity. While this was latent in the overall business case, it had never been defined physically or translated into a budget. the architects did that.

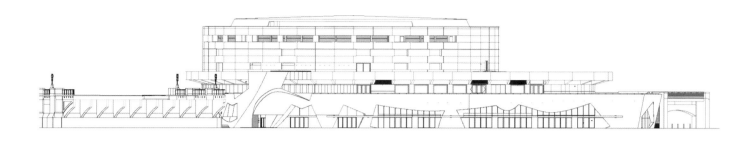

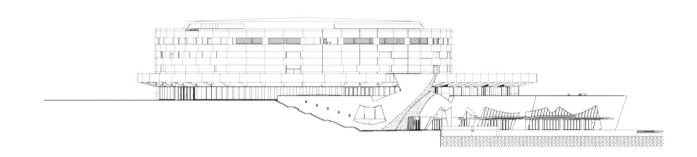

Elevation 1

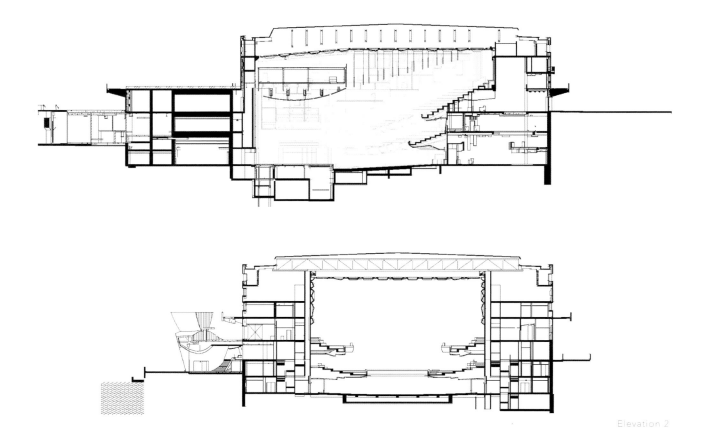

ARM coordinated the design team which included Peter Elliott (Urban Design), TCL (Landscape), Kirkegaard Associates and Marshall Day (acousticians), Aurecon (Services & Structure) and Shuler Shook (theatre designers).

The project was delivered through an Alliance contract, which means that the architects were an equal member of the project delivery team comprising Owner, Constructor and Designer. Hamer Hall reborn was delivered within the State's target budget, and within the prescribed but unrealistically short time frame. Anecdotally, the scope and functionality have been achieved well beyond "business as usual" and the expectations of the client.

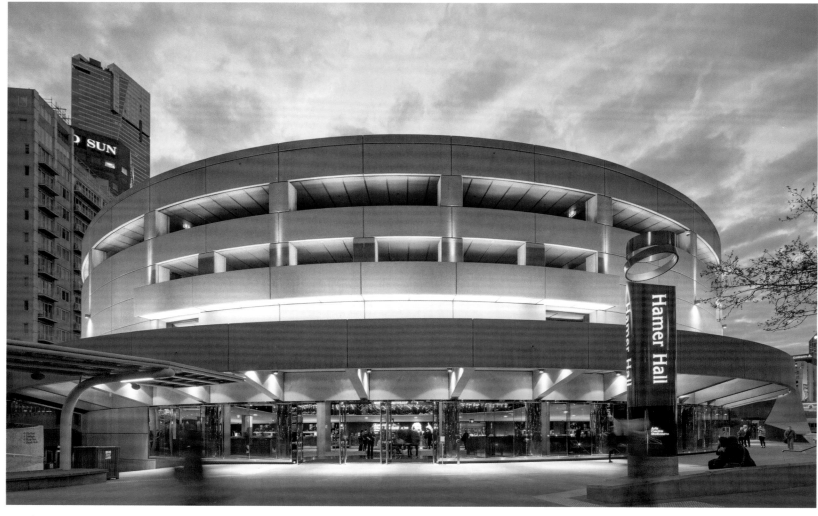

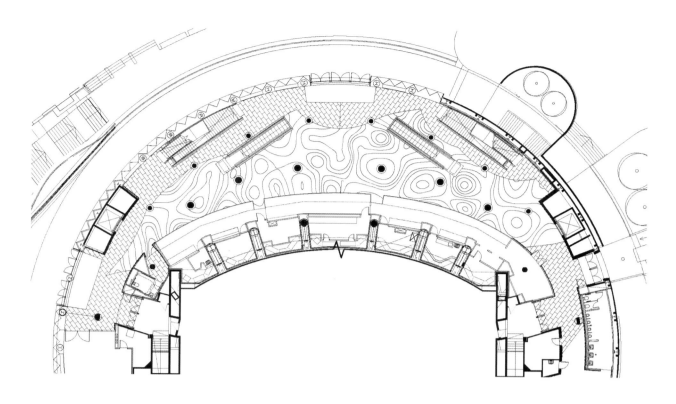

Detail 1

Detail 2

253

Detail 3

255

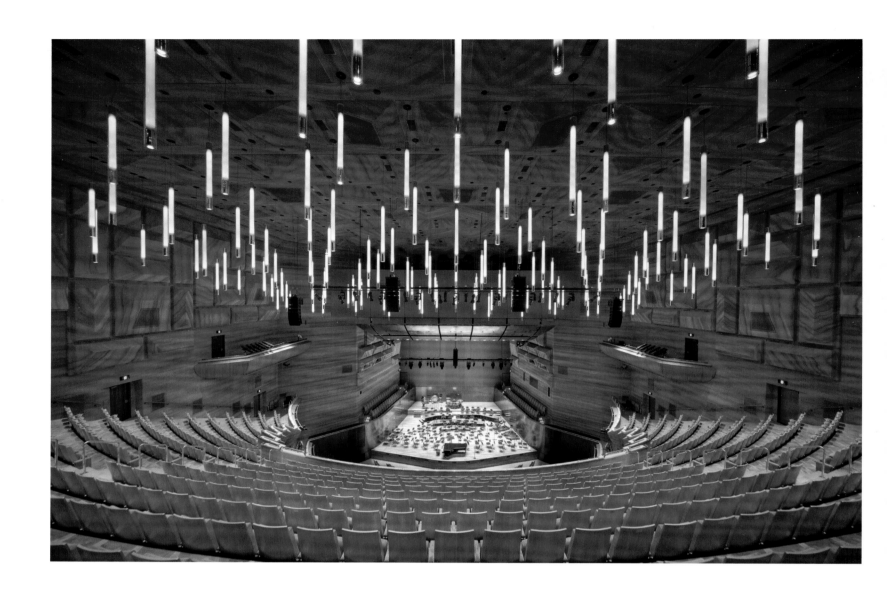

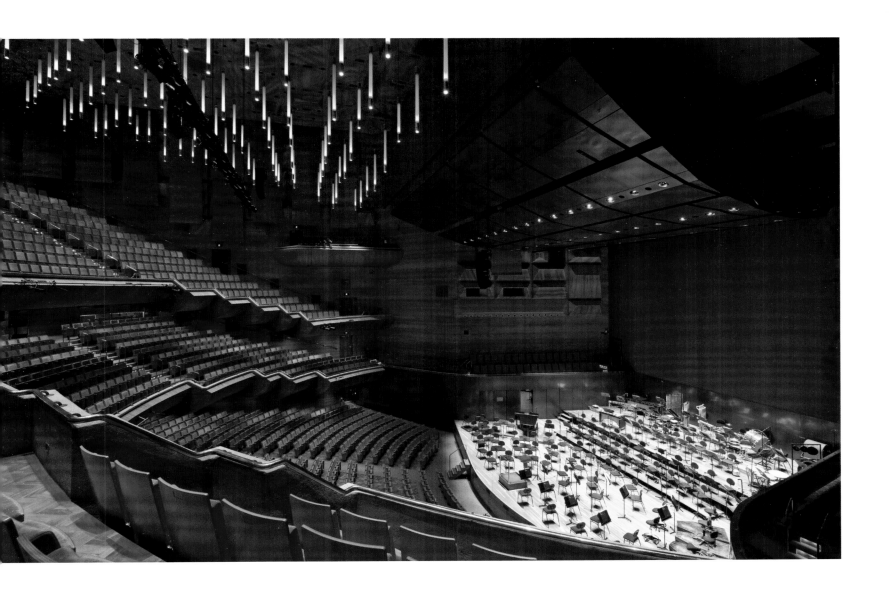

Keywords

Dynamic Forms
Triangular Tensions
Attractive Skins

PALOMA
Contemporary music venue at Nîmes Métropole

Location: Nîmes, France
Architectual Design: Tetrarc Architects
Project Director: Michel Bertreux
Project Manager: Rémi Tymen
Study: Olivier Perocheau, Richard Sicard, Florent Delaboudinière, Timothée Naux
Landscape Designer: Louise Follin
Construction Site: Guillaume Blanchard, Marc-Antoine Bouyer, Alice Pedel
Computer Graphics Designer: Mikaël Trocmé
Area: 5,611 m^2
Photography: Stéphane Chalmeau

Situated in southern France, Nîmes is well known for its exuberant character, its fashionable artists and its tumultuous/spectacular storms.

Tetrarc has fed on Nîmes' special identity in order to create the complex dedicated to contemporary music that has just been opened at the entrance to the town between a flying club and a district made up of a mix of low rise buildings, houses and small local businesses.

Given the name Paloma, the complex consists of two concert halls, seven rehearsal and recording studios, six accommodation areas for the performers in residence, administration offices and the all-important technical facilities.

Tetrarc has devised a tonic form which springs out from the ground and flows out towards the sky and the town as if some powerful internal forces are pushing at its walls and are threatening to shatter them. This zinc shell actually stretches, frays, even tears apart in some parts in order to display– like a supernatural eye – the giant screen which announces performances and artists.

This is also an internal event. Viewing the concert as a confrontation between artists and the public, Tetrarc uses the colors of the bullfight (yellow and purple) for the foyer, the stairs and the patio; red in the foyers at the entrance to the two halls; the geometry of the bullfighter's movements are represented by the congealed textures on the walls of the hall; images of a crowd seated in an arena are projected onto the seats in the big hall. The walls of this hall display a giant sculpture whose material evokes the sleekness of the picadors' hair, and whose form evokes a gigantic cog like the one grinding the men in Charlie Chaplin's Modern Times.

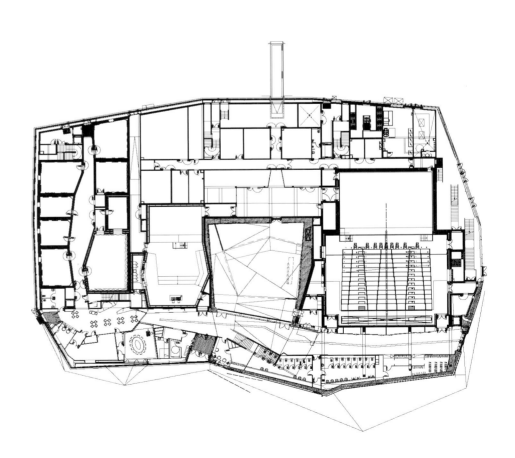

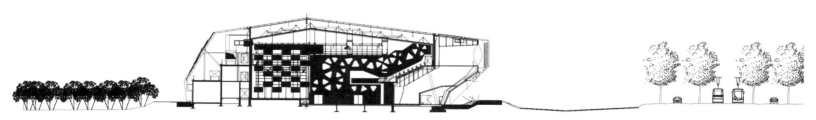

262 | Art Museum

For Tetrarc, the architecture is at the confluence of artistic expression. There is an echo of the cinema of Spielberg and Chaplin. The visual art is a feature of the foyer conceived like a penetrable sculpture opened out into the interior space of the shell as well as the small hall, the studios and the performers' restaurant which plays tribute to abstract geometric art, whilst the big repetitive stripes on the patio evoke the work of the Support-Surface group, a well known presence in Nîmes. The design is linked to the music roots evident in the choice of vintage furnishings redolent of the 1960s in the restaurant and the performers' apartments.

Culturally rich, Tetrarc's project has also been retained in the precision of its planning: the different entrances (public, administration, performers, materials) are clearly separated; the halls are served directly from the vast foyer opening onto the patio, the sales area for associated products, the cloakroom and the radio studio; the three stages are directly connected and on the level of a single loading bay for the lorries that supply the equipments... and the activity on the floor of the immense concrete square draws the eyes of the storm towards the big retention pond. After The Factory in Nantes, Tetrarc is diving into its rocker past to make Paloma rock.